THE OXFORD BOOK OF AGES

THE
OXFORD BOOK OF
AGES

CHOSEN BY
ANTHONY AND SALLY SAMPSON

Oxford   New York
OXFORD UNIVERSITY PRESS
1985

Oxford University Press, Walton Street, Oxford OX2 6DP

London New York Toronto
Delhi Bombay Calcutta Madras Karachi
Kuala Lumpur Singapore Hong Kong Tokyo
Nairobi Dar es Salaam Cape Town
Melbourne Auckland
and associates in
Beirut Berlin Ibadan Mexico City Nicosia

Oxford is a trade mark of Oxford University Press

British Library Cataloguing in Publication Data
The Oxford book of ages.
1. Aging
I. Sampson, Anthony   II. Sampson, Sally
612'.67   QP86
ISBN 0–19–214134–1

Library of Congress Cataloging in Publication Data
Main entry under title:
The Oxford book of ages.
Includes index.
1. Age—Quotations, maxims, etc. 2. Life cycle,
Human—Quotations, maxims, etc. I. Sampson, Anthony.
II. Sampson, Sally.
PN6084.A3509 1985   082   84-20752
ISBN 0-19-214134-1

Set by Rowland Phototypesetting Ltd
Printed in Great Britain by
Richard Clay (The Chaucer Press) Ltd.
Bungay, Suffolk

CONTENTS

Introduction   vii

THE OXFORD BOOK OF AGES   1

Acknowledgements   187

Index of Names   197

INTRODUCTION

❧

THIS anthology is meant to speak for itself, as a selection of reflections about each year of life, from birth to a hundred. It is essentially a personal choice: we have not sought to underline particular themes, but to pick out quotations which appeal to us with their originality, variety, and interest. By finding quotations for each year and omitting more generalized reflections—which are plentiful in other anthologies—we have tried not only to give a sense of continuity and gradual development, but to cut down the kind of self-conscious theorizing which so often overcomes writers when faced with the generalities of middle age, or with the traditional ordeals of facing up to birthdays in each new decade. These ordeals of course have their own significance, and they naturally call for ritual lamentations; but the less formal and more intimate thoughts of letter-writers or diarists about the passage of years are often more enlightening, because less studied.

We have looked for a wide range of authors, including American and Continental writers, though without trying to do justice to Oriental concepts of ageing and wisdom which reflect a different philosophical perspective. Among scores of contributors we have tried to find a balance between pessimism and optimism, between average and exceptional achievements, between late and premature ageing. But we have also given special attention to a few individual writers—ranging from Cicero to Emerson, from Samuel Johnson to Ogden Nash—who have been unusually articulate and self-aware in writing about their advancing years, and who provide a kind of recurring chorus against the changing background of voices. From the age of eighty upwards our choice is obviously more restricted, and much of the available material is melancholy and repetitive. We have concentrated attention on exceptional old people, such as Bernard Berenson or Bernard Shaw, who without being representative provide outspoken views.

No literary anthology can hope to be typical of the attitudes and feelings of ordinary people. Creative minds have their own obsessions about ageing, about passing their prime and losing their inspiration, and they are less likely to grow old quietly. The writers, painters and musicians who contribute most to this collection operate outside the normal milestones of an ordered career, or annual increments, promotions or fixed retirements. The different arts, moreover, make very different demands. Poets—with mathematicians—have always been seen as early developers, who are thus especially worried about losing their originality in middle age. Wordsworth insisted that 'a poet who has not produced a good poem before he is twenty-five, we may conclude cannot, and never will do so'—though Thomas Hardy did not publish his first book of poems till the age of fifty-eight. Novelists, playwrights, and painters may often produce their first important work well past their twenties, and some philosophers—including John Locke and Immanuel Kant —have produced their first major work in their fifties. Musicians on the other hand can appear as prodigies of both youth and old age: they must reveal their talent early, and they need an intensive training, yet some of them—such as Verdi, Sibelius, and Casals —continue composing or performing into old age. Conductors, who are compelled to take vigorous exercise, are notable for their longevity; and the performing arts in general, which demand intense mental and physical discipline, provide many of the most remarkable examples of active old age—with actors among the most lively. Writers, following a comparatively sedentary and lonely profession, appear more vulnerable to physical or psychological decline.

Certainly theories about the need for conservation of energy find little encouragement from these quotations: the most active people who face recurring challenges often enjoy the most vigorous old age. The importance of challenge and will-power is evident everywhere. Artists, politicians, or entrepreneurs do not have to endure the sudden shock and humiliation of forcible retirement in their sixties, which so often induces a premature slowing-down; and many of the most durable old people, like Sir Robert Mayer at 100, are determined to 'die in harness'. 'If I stopped writing', wrote Bernard Shaw at 91, 'I should die for want of something to do.' Among public figures the need for a harness is much more obvious, and ageing politicians like Wellington, Gladstone or Churchill, appear to be kept going by the challenges and disciplines of office.

But creative genius is more often associated with early death; and many of the great prodigies, like Mozart or Byron, have shown a presentiment of an early death in the frenzied pace of their lives and art. Their concentrated achievement remains a puzzle to those who survive them, who cannot imagine them growing old; Goethe reflected how Mozart, Raphael, and Byron—all dead in their thirties —'had perfectly fulfilled their missions; and it was time for them to depart, that other people might still have something to do in a world made to last a long while'. There is obvious support for Schopenhauer's observation:

. . . the character of almost every man seems to be pre-eminently adapted to *one* stage of life; so that in this stage he appears to the greatest advantage. Some are lovable youths, and that is all; others are active and energetic men, but age robs them of all worth; many appear most advantageously in old age, when they are more benevolent, because more experienced and more tranquil. The reason for this must be that the character itself has something youthful, or adult, or elderly about it, with which quality the current stage of his life harmonizes, or opposes as a corrective.

Many of the quotations in this book seem to encourage this concept; yet there are striking examples of people—like Hardy and Yeats among writers or Churchill and Gladstone among public men —who can achieve a second or third flowering at later ages, revealing different characteristics. And some of the most vigorous lives, like Churchill's, have been able to embrace different careers—whether a soldier's, a journalist's, or a politician's—each at a suitable stage.

Certainly the gloomy forebodings of Byron or Dylan Thomas in their twenties are not typical. For most writers, as for most ordinary people, the concept of mortality does not loom large until the mid-twenties or early thirties: Siegfried Sassoon at twenty-two thought himself inextinguishable; 'the young man till thirty', says Charles Lamb, 'never feels practically that he is mortal'. Many people who have been terrified in their teens like Simone de Beauvoir by 'the thought of that utter non-being', put it out of their minds until much later. It is not till the thirties that the thoughts of failure, decline, and a limited time-span begin to crowd in, and when the more self-indulgent become aware of the damage done by their excesses. But forebodings about failure do not necessarily correspond to the future in store. There are enough late developers, among novelists, politicians, or philosophers, to give some hope to the unfulfilled, and a few people during their thirties strike out in a new career, like Gauguin as a painter, or Schweitzer as a doctor in Africa. It is the fortieth birthday, a favourite subject for aphorisms, which provides the classic occasion for reassessment, regrets, and complaints about the end of youth.

The later forties and fifties provide the leanest period for comments and introspection about ageing. They offer no surprises in terms of prodigious early development or remarkable longevity; and they are part of the wide plateau of middle age, on to which people have climbed without yet having to descend, whose precipices may be out of sight. Many artists and public men find themselves too preoccupied with their own worldly challenges to want to think too much about the coming years of decline.

Obviously attitudes to ageing and survival have changed fundamentally over the last few centuries, which form the background to most of this anthology. The low expectation of life and the commonness of infant mortality in Elizabethan England provided a quite different perspective to death and longevity. The biblical reference to a life of three score years and ten was far above the average life-span at the time, and in Roman times Cicero reckoned that 'old age begins at forty-six years, according to the common opinion'. The age of thirty-three, at which Christ was believed to have died, marked a traditional milestone for men to measure their own achievement, while the 'grand climacteric' of sixty-three was regarded up to the seventeenth century as a year of special danger. Yet every era has had its small quota of distinguished octogenarians and nonagenarians who were all the more phenomenal against a background of short life-expectation, health hazards and plagues. While Thomas Hobbes condemned human life as 'nasty, brutish and short' he himself was able to write a passable poem at ninety.

The whole concept of childhood as a distinct and separate age, with its own formative or romantic significance, is a very recent development. In the Middle Ages, at a time of child-kings and teenage generals, there was little scope for idealizing children. But by the early nineteenth century the belief that children had their own kind of wisdom and understanding, 'trailing clouds of glory', hugely affected attitudes towards them not only among poets but among philosophers and scientists: 'may we not suspect', wrote Darwin in 1877, 'that the vague but very real fears of children, which are quite independent of experience, are the inherited effects of real dangers and abject superstitions during ancient savage times?' The conflict between different approaches to children remains, and runs through the earlier pages of this anthology. Few writers can remember their own thoughts and opinions before the age of six; and in covering these early years we try to contrast the romantic hindsights of poets and novelists with the more severe judgements of philosophers and psychologists. The Romantics changed attitudes not only to childhood, but to the whole measurement of time. Before the eighteenth century there was little written self-analysis and introspection about the passing years of the kind which writers from Rousseau onwards loved to put into their letters, diaries, and autobiographies.

This is not a medical or scientific anthology, and it carries only incidental references to the biological or hereditary causes of ageing.

Such evidence as emerges about the importance of diet, or temperance, is thoroughly confusing and contradictory. The exasperating Luigi Cornaro, the author of *An Earnest Exhortation to a Sober Life* in the sixteenth century, lived to be 100 by following the most fastidious diet; while many active octogenarians like Bertrand Russell or Sibelius claimed to have taken no trouble with their food: 'I have always lived an unhealthy life', said Sibelius at eighty-seven. Nor is there any agreement about psychological recipes for old age; there are glimpses of serenity, but many pictures of restless, combative, or agonized old people. 'My eighties are passionate,' wrote Florida Scott-Maxwell at eighty-two. 'I grow more intense with age'; and E. M. Forster was aware at the same age of a proneness to 'senile lechery'. Bertrand Russell contradicted any thoughts of serenity, and the great Japanese painter Hokusai liked to call himself 'the old man, mad about drawing'. W. B. Yeats asked: 'Why should not old men be mad?'

Most writers and poets, from Wordsworth to Auden, seem to follow the same political development as most politicians and voters, moving from radicalism in their youth to conservatism in middle age, endorsing the precept of Aristide Briand: 'The man who is not a socialist at twenty has no heart, but if he is still a socialist at forty he has no head.' Robert Louis Stevenson, already aware of his growing conservatism at thirty one, submitted to it 'as I would submit to gout or grey hair as a concomitant of growing age or else of failing animal heat; but I dare say it is deplorably for the worst'. The change of heart embarrasses many writers—particularly the most revolutionary—when they look back on their earlier work; while others are determined to avoid such embarrassment. As Robert Frost wrote in his poem 'Precaution':

> I never dared to be radical when young
> For fear it would make me conservative when old.

But there are also writers and politicians, like Bertrand Russell or William Gladstone, who move in the opposite direction, towards radical or revolutionary principles, and who seem to gain extra energy and challenge from their mounting sense of indignation or outrage.

The search for happiness provides a more plaintive theme as the years advance; yet there are enough examples of finding contentment in middle or later age to offset some of the gloom from writers who feel they are losing it as their powers diminish. Balzac insists at the age of thirty-six that 'if at my age one has still not known happiness, pure and unalloyed, Nature will not henceforth allow one to put the cup to one's lips'. But other writers begin to feel both happier and younger in later years, like Henry Miller who said 'it was only in the forties that I started feeling young'; and a few who belatedly find sexual or psychological fulfilment: 'I have drunk of the wine of life at last', wrote Edith Wharton at forty-six when she first fell in love; 'From fifty-one to fifty-three I have been happy,' wrote E. M. Forster, 'and would like to remind others that their turn can come too.' C. S. Lewis looked forward to the happiness in his sixties which he missed in his twenties. 'I've discovered too late', said Lord Reith at seventy-eight, 'that life is for living.' Raymond Mortimer claimed that the five years after seventy were the happiest in his life. The pianist Artur Rubinstein, who liked to say that he was the happiest man he had met, reaffirmed it at ninety-one.

The traditional milestone of seventy—which now corresponds roughly to the average expectation of life in Western countries —provides the almost inescapable opportunity for rethinking attitudes to life. Septuagenarians begin to feel that they are living on borrowed time, or enjoying a bonus, liberated from past responsibilities or constraints. 'If I don't enjoy myself now,' wrote Omar Khayyám, 'when shall I?' But those who retain their creative drive can never decide whether they want to retire from the contest, or find new challenges. 'My remaining days I may now consider a free gift,' wrote Goethe at eighty-one; 'and it is now, in fact, of little consequence what I now do, or whether I do anything'; yet only a year later he was delighted to find that 'even at my great age ideas come to me the pursuit and development of which would require another lifetime'. Many artists and writers over seventy still like to pace themselves by recalling old men of earlier days. Longfellow at eighty looked back to Cato and Goethe: 'Ah, nothing is too late/ Till the tired heart shall cease to palpitate.' And Thomas Hardy recalled Sophocles and Homer, who 'Burnt brightlier towards their setting-day'.

But all elderly people, however brightly burning, have to agree about many of their ordeals—not just the weakening faculties, limbs, and memories, but the changed perspective of death and the sense of time passing constantly faster. After sixty, says Don Marquis, 'in ten minutes more you are eighty-five'; or as the nineteenth-century poet Thomas Campbell puts it:

> The more we live, more brief appears
> Our life's succeeding stages;
> A day to childhood seems a year,
> And years like passing ages.

The passing seasons and the patterns of nature loom larger; so that Sibelius at eighty-nine could say that the change of seasons is 'the most important thing in my life'. And older people inevitably become more self-conscious of their age as it separates them from most of the population; while they become prouder of their age, often preferring (as they did in their teens) to be thought older than they are. The sense of isolation dominates the writing of most over-eighties as they survey the younger generations: 'as though men spend their lives perched upon living stilts which never cease to grow'—as Proust described the Duc de Guermantes at eighty-three. The tiny handful of centenarians, looking across the gap of an extra generation, can never forget that they stand (as Margaret Murray described it) 'on a high peak alone'.

Yet old people, particularly grandparents and great-grandparents, can often establish their own natural relationships with the young, both sides feeling themselves outside the rational, competitive world of the intervening generations. They can both silently say to each other (as John Cowper Powys wrote at eighty-four): 'Lord! what fools these grown-ups be!' The sevens and the seventy-fives, as J. B. Priestley expresses it, 'are both closer to the world of mythology and magic than all the busier people between those ages'. Finding friendships and common interests with the young can give added confidence to artists as well as to ordinary people; while the idea of rallying the young against the conformity and conservatism of middle age has inspired ageing leaders as different as Mao Tse-Tung and Gladstone, who proclaimed at eighty-three: 'I represent the youth and hope of England.' The tricks of memory in old age conspire to bridge the long divide to childhood, as old people find they can remember events seventy years ago much more easily than the happenings of the previous week. The consciousness of a second childhood, in a more interesting sense than senility or dependence, looms larger in the final years.

Finally, a note on the disposition of quotations. This is obvious where an age is mentioned in the course of the piece; where this is not the case, the author or subject was at that age when the reflection was made. No attempt has been made within a year to impose a historical or any other framework; the juxtapositions are sometimes meant to be illuminating, sometimes amusing, and generally point to the great diversity of experience of people at the same age. It has also to be remembered that when someone is 'in his sixtieth year', he is in fact· fifty-nine years old, not sixty: this accounts for apparently different 'ages' occasionally falling under the same heading.

ANTHONY AND SALLY SAMPSON

Our birth is nothing but our death begun.

EDWARD YOUNG, Night Thoughts, 1742–6

There are but three events in a man's life: birth, life and death. He is not conscious of being born, he dies in pain, and he forgets to live.

JEAN DE LA BRUYÈRE, Characters, 1688

> Birth, and copulation, and death.
> That's all the facts when you come to brass tacks:
> Birth, and copulation, and death.
> I've been born, and once is enough.
> You don't remember, but I remember,
> Once is enough.

T. S. ELIOT, Sweeney Agonistes, 1932

The cradle rocks above an abyss, and common sense tell us that our existence is but a brief crack of light between two eternities of darkness. Although the two are identical twins, man, as a rule, views the pre-natal abyss with more calm than the one he is heading for (at some 45 hundred heartbeats an hour).

VLADIMIR NABOKOV, Speak, Memory. An Autobiography Revisited, 1967

> These little limbs,
> These eyes and hands which here I find,
> This panting heart wherewith my life begins;
> Where have ye been? Behind
> What curtain were ye from me hid so long!
> Where was, in what abyss, my new-made tongue?

THOMAS TRAHERNE (1638–74), 'The Salutation'

The period following birth must be regarded as the time when the nourishment given to a child has the greatest effect on the development of the body.

ARISTOTLE, *Politics, c.* 345 BC

It is the physical weakness of a baby that makes it seem 'innocent', not the quality of its inner life. I myself have seen a baby jealous; it was too young to speak, but it was livid with anger as it watched another baby at the breast.

ST AUGUSTINE, *Confessions,* I. vii, 397–401

Even as therefore in generation the body goeth before the soul, so doth the unreasonable part of the soul go before the reasonable: the which is plainly to be discerned in young babes, who (in a manner) immediately after their birth utter anger and fervent appetite, but afterward in process of time reason appeareth.

CASTIGLIONE, *The Courtier,* 1528

This tender age is like water spilt upon a table, which with a finger we may draw and direct which way we list; or like the young hop, which, if wanting a pole, taketh hold of the next hedge; so that now is the time (as wax) to work it pliant to any form.

HENRY PEACHAM, *The Compleat Gentleman,* 1622

I think we may observe, that when children are first born, all objects of sight, that do not hurt the eyes, are indifferent to them; and they are no more afraid of a blackamoor, or a lion, than of their nurse or a cat. What is it then, that afterwards, in certain mixtures of shape and colour, comes to affright them? Nothing but the apprehensions of harm, that accompany these things. Did a child suck every day a new nurse, I make account it would be no more affrighted with the changes of faces at six months old than at sixty.

JOHN LOCKE, *Some Thoughts Concerning Education,* 1693

              'I have no name:
              I am but two days old.'
              What shall I call thee?
              'I happy am,
              Joy is my name.'
              Sweet joy befall thee!

Pretty Joy!
Sweet Joy, but two days old,
Sweet Joy, I call thee:
Thou dost smile,
I sing the while,
Sweet joy befall thee!

WILLIAM BLAKE, 'Infant Joy', *Songs of Innocence*, 1789

Johnson was pleased with my daughter Veronica then a child of about four months old. She had the appearance of listening to him. His motions seemed to her to be intended for her amusement; and when he stopped she fluttered, and made a little infantile noise, and a kind of signal for him to begin again. She would be held close to him: which was a proof from simple nature that his face was not horrid. Her fondness for him endeared her still more to me, and I declared she should have £500 of additional fortune.

JAMES BOSWELL, *Journal of a Tour to the Hebrides*, 1785

An ugly baby is a very nasty object—and the prettiest is frightful when undressed—till about four months; in short as long as they have their big body and little limbs and that terrible frog-like action.

QUEEN VICTORIA to the Princess Royal, 2 May 1859

The right moment to begin the requisite moral training is the moment of birth, because then it can be begun without disappointing expectations.

BERTRAND RUSSELL, *On Education, especially in Early Childhood*, 1926

The character of a child is already plain, even in its mother's womb. Before I was born my mother was in great agony of spirit and in a tragic situation. She could take no food except iced oysters and champagne. If people ask me when I began to dance I reply, 'In my mother's womb, probably as a result of the oysters and champagne —the food of Aphrodite'.

ISADORA DUNCAN, *My Life*, 1927

Psychoanalysts who specialize in the study of children have no doubt as to the extreme aggressiveness of the frustrated infant.

ANTHONY STORR, *Human Destructiveness*, 1972

And underneath the pram cover lies my brother Jake
He is not old enough yet to be properly awake
He is alone in his sleep; no arrangement they make
For him can touch him at all, he is alone,
For a little while yet, it is as if he had not been born
Rest in infancy, brother Jake; childhood and interruption come
    swiftly on.

STEVIE SMITH, 'Childhood and Interruption'

1

Any one may observe that from the fifth or sixth month children employ their whole time for two years and more in making physical experiments. No animal, not even the cat or dog, makes this constant study of all bodies within its reach; all day long the child of whom I speak (at twelve months) touches, feels, turns round, lets drop, tastes and experiments upon everything she gets hold of; whatever it may be, doll, ball, coral or plaything, when once it is sufficiently known she throws it aside, it is no longer new, she has nothing to learn from it and has no further interest in it. It is pure curiosity; physical need, greediness, count for nothing in the case; it seems as if already in her little brain every group of perceptions was tending to complete itself, as in that of a child who makes use of language.

HIPPOLYTE TAINE, *On the Acquisition of Language by Children*, 1876

Jealousy was plainly exhibited when I fondled a large doll, and when I weighed his infant sister, he being then fifteen and a half months old. Seeing how strong a feeling jealousy is in dogs, it would probably be exhibited by infants at any earlier age than that just specified, if they were tried in a fitting manner.

CHARLES DARWIN, *A Biographical Sketch of an Infant*, 1877

The first memory of my life dates from my second year, when a maid dropped me against the chimney-piece. I was cut on the forehead and frightened. This shock jolted me into awareness of life, and I clearly saw, I still see, the rosy marble of the chimney-piece, and my blood running down it, and the distraught look of the maid. I also remember the doctor's visit, the leeches he applied behind my ear, my mother's anxiety, and the maid's dismissal for drunkenness. We left that house, wherever it was; I've never been back, but I feel that I'd immediately find my way about it.

GEORGE SAND, *My Life*, 1854

It was a peculiarity of this baby to be always cutting teeth. Whether they never came, or whether they came and went away again, is not in evidence; but it had certainly cut enough, on the showing of Mrs Tetterby, to make a handsome dental provision for the sign of the Bull and Mouth . . . Mrs Tetterby always said, 'it was coming through, and then the child would be herself;' and still it never did come through, and the child continued to be somebody else.

CHARLES DICKENS, 'The Haunted Man', 1848

There he lay upon his back,
The yearling creature, warm and moist with life
To the bottom of his dimples,—to the ends
Of the lovely tumbled curls about his face;
For since he had been covered over-much
To keep him from the light-glare, both his cheeks
Were hot and scarlet as the first live rose
The shepherd's heart-blood ebbed away into
The faster for his love. And love was here
As instant; in the pretty baby-mouth,
Shut close as if for dreaming that it sucked,
The little naked feet, drawn up the way
Of nestled birdlings; everything so soft
And tender,—to the tiny holdfast hands,
Which, closing on a finger into sleep,
Had kept the mould of 't.

ELIZABETH BARRETT BROWNING, *Aurora Leigh*, 1857

I am all bound up; I want to stretch out my arms and I cannot. I scream and cry and I hate my own screaming, but I cannot stop. People are leaning over me—I can't remember who—and everything is shrouded in semi-darkness. There are two of them. My screaming affects them; they are anxious; but they do not release me as I want them to, and I scream still louder.

TOLSTOY, recollecting his life before his mother died, when he was two

In a child of about one year old the anxiety caused by the beginning of the Oedipus conflict takes the form of a dread of being devoured and destroyed.

MELANIE KLEIN, *Early Stages of the Oedipus Conflict*, 1928

At one year of age the child says his first *intentional* word. He babbles as before, but his babbling has a purpose, and this intention is a proof of conscious intelligence.

DR MARIA MONTESSORI, *The Absorbent Mind*, 1949

He was given a primer 'reading made easy', known as a 'tipenny', and he is reported to have known his letters by the time he was eighteen months old.

JAMES MURRAY, editor of the Oxford English Dictionary, described by K. M. Elisabeth Murray in *Caught in the Web of Words: James Murray and the Oxford English Dictionary*, 1977

2

A marked change occurs when the child develops consciousness of his ego, a fact which is registered by his referring to himself as 'I'. This change normally takes place between the third and fifth year, but it may begin earlier.

C. G. JUNG, *The Development of Personality*, 1934

In the period between two and three, children are apt to show signs of contrariness and other inner tensions . . . The one-year-old contradicts his mother. The two-and-a-half-year-old even contradicts himself . . . It is often hard to get along with a child between two and three.

DR BENJAMIN SPOCK, *Baby and Child Care*, 1955

Four stages can be picked out of the evolution of modality. The first lasts till the age of two to three, the second extends from two to three to seven to eight, the third from seven to eight to eleven to twelve, and the fourth begins at this age. During the first stage, reality may be said to be simply and solely what is desired . . . The second stage marks the appearance of two heterogeneous but equal realities—the world of play and the world of observation. The third marks the beginning of hierarchical arrangement, and the fourth marks the completion of this hierarchy, thanks to the introduction of a new plane—that of formal thought and logical assumptions.

> JEAN PIAGET, *Judgement and Reasoning in the Child*, 1926

A little later (two and half years) she was very much struck by the sight of the moon. She wanted to see it every evening; when she saw it through the window-panes there were cries of joy; when she walked it seemed to her that it walked too, and this discovery charmed her . . . All this closely resembles the emotions and conjectures of primitive peoples, their lively and deep admiration for natural objects, the power that analogy, language and metaphor exercise over them, leading them to solar and lunar myths.

> HIPPOLYTE TAINE, *On the Acquisition of Language by Children*, 1876

It is well-known how intensely older children suffer from vague and undefined fears, as from the dark, or in passing an obscure corner in a large hall, etc. I may give as an instance that I took the child in question, when two and a half years old, to the Zoological Gardens, and he enjoyed looking at all the animals which were like those he knew, such as deer, antelopes, etc., and all the birds, even the ostriches, but was much alarmed at the various larger animals in cages. He often said afterwards that he wished to go again, but not to see 'beasts in the houses'; and we could in no manner account for this fear. May we not suspect that the vague but very real fears of children, which are quite independent of experience, are the inherited effects of real dangers and abject superstitions during ancient savage times?

> CHARLES DARWIN, *A Biographical Sketch of an Infant*, 1877

At two year and half old he could perfectly read any of the English, Latin, French or Gothic letters; pronouncing the three first languages exactly.

> JOHN EVELYN, about his son Dick. *Diary*, 1658

My feelings were very acute; they used to amuse themselves by making me cry at sad songs and dismal stories. I remember 'Death and the Lady', 'Billy Pringle's Pig', 'The children sliding on the ice all on a summer's day', and Witherington fighting on his stumps at

Chevy Chase. This was at two years old, when my recollection begins—prior identity, I have none—they tell me I used to beg them not to proceed.

> ROBERT SOUTHEY, letter to G. C. Bedford, 30 September, 1796

My earliest memory of all is a mad passion of rage at my elder brother John (on a visit to us likely from his grandfather's); in which my Father too figures though dimly, as a kind of cheerful comforter and soother. I had broken my little brown stool, by madly throwing it at my brother; and felt for perhaps the first time, the united pangs of Loss and of Remorse. I was perhaps hardly more than two years old; but can get no one to fix the date for me, though all is still quite legible for myself.

> CARLYLE, *Reminiscences*, 1881

Half in fun, half in earnest, I learned to know the keys by their names, and with my back to the piano I would call the notes of any chord, even the most dissonant one. From then on it became 'mere child's play' to master the intricacies of the keyboard.

> ARTUR RUBINSTEIN, *My Young Years*, 1973

At the age of two, I apparently did a passable imitation of Lloyd George.

> PETER USTINOV, *Dear Me*, 1977

3

Loving she is, and tractable, though wild;
And Innocence hath privilege in her
To dignify arch looks and laughing eyes
And feats of cunning; and the pretty round
Of trespasses, affected to provoke
Mock-chastisement and partnership in play.
And, as a faggot sparkles on the hearth,
Not less if unattended and alone
Than when both young and old sit gathered round

And take delight in its activity;
Even so this happy Creature of herself
Is all-sufficient; solitude to her
Is blithe society, who fills the air
With gladness and involuntary songs.
Light are her sallies as the tripping fawn's
Forth-startled from the fern where she lay couched;
Unthought-of, unexpected, as the stir
Of the soft breeze ruffling the meadow-flowers,
Or from before it chasing wantonly
The many-coloured images imprest
Upon the bosom of a placid lake.

> WORDSWORTH, 'Characteristics of a Child Three Years Old',
> 1815

Thou pretty opening rose!
(Go to your mother, child, and wipe your nose!)
Balmy and breathing music like the South,
(He really brings my heart into my mouth!)
Fresh as the morn, and brilliant as its star, —
(I wish that window had an iron bar!)
Bold as the hawk, yet gentle as the dove, —
(I'll tell you what, my love,
I cannot write, unless he's sent above!)

> THOMAS HOOD, 'A Parental Ode to my Son, Aged Three
> Years and Five Months', 1837

I have heard the parents of a very large family of different characters
declare that instinctive obedience can be, as a general rule, learnt by
three-year-olds.

> CHARLOTTE M. YONGE, 'The Parents' Power', address to the
> Conference of the Mothers' Union, 1889

James James
Morrison Morrison
Weatherby George Dupree
Took great
Care of his Mother
Though he was only three.

> A. A. MILNE, 'Disobedience', *When We Were Very Young*, 1924

When a healthy three-year-old child says 'I love you' there is meaning in it like that between men and women who love and are in love.

D. W. WINNICOTT, *The Child, the Family, and the Outside World*, 1964

Ordinarily, we do not expect a child to show any fear of the dark until he has reached the age of three.

DOUGLAS A. THOM, MD, *Everyday Problems of the Everyday Child*, 1937

I was dreadfully alive to nervous terrors. The night-time solitude, and the dark, were my hell. The sufferings I endured in this nature would justify the expression. I never laid my head on my pillow, I suppose, from the fourth to the seventh or eighth year of my life—so far as memory serves in things so long ago—without an assurance, which realized its own prophecy, of seeing some frightful spectre.

CHARLES LAMB, *Witches and Other Fears*, 1821

One day, when I was three and a half years old, I was standing next to my father, as he played a Mozart string quartet with his friends. It began with the notes D.B.G. 'How do you know you must play these three notes?', I asked him. Patiently he took a sheet of paper, drew the five lines of a musical staff, and explained what each note meant when written between or on given lines . . . I understood at once what he was trying to teach me. And so it came about that I literally could reach music before I learnt my ABC!

FRITZ KREISLER, the violinist

What folly it is that daughters are always supposed to be
In love with Papa. It wasn't the case with me
I couldn't take to him at all
But he took to me.
What a sad case to befall
A child of three.

STEVIE SMITH, 'Papa Love Baby'

John Grubby, who was short and stout
And troubled with religious doubt,
Refused about the age of three
To sit upon the curate's knee.

G. K. CHESTERTON (1874–1936), *The New Freethinker*

4

It has been found that in early childhood there are signs of bodily activity to which only an ancient prejudice could deny the name of sexual, and which are linked to psychical phenomena that we come across later in adult erotic life—such as fixation to particular objects, jealousy and so on. It is further found, however, that these phenomena which emerge in early childhood form part of an ordered course of development, that they pass through a regular process of increase, reaching a climax towards the end of the fifth year, after which there follows a lull. During this lull progress is at a standstill and much is unlearnt and there is much recession. After the end of this period of latency, as it is called, sexual life advances once more with puberty; we might say that it has a second efflorescence. And here we come upon the fact that sexual life is *diphasic*, that it occurs in two waves—something that is unknown except in man.

SIGMUND FREUD, *An Outline of Psycho-Analysis*, 1949

While I'm emphasizing how agreeable children usually are between three and six, I ought to make a partial exception for four-year-olds. There's a lot of assertiveness, cockiness, loud talk and provoking that comes out around four years in many children and that requires a firm hand in the mother.

DR BENJAMIN SPOCK, *Baby and Child Care*, 1955

A remarkably intelligent little girl of four years, but who had never in her own family been used to the common phrases which sometimes pass for humour, happened to hear a gentleman say, as he looked out of the window one rainy morning, 'It rains cats and dogs today'. The child, with a surprised but believing countenance, immediately went to look out at the window to see the phenomenon.

MARIA AND R. L. EDGEWORTH, *Essays on Practical Education*, 1822

The first sense of sorrow I ever knew was upon the death of my father, at which time I was not quite five years of age; but was rather amazed at what all the house meant, than possessed with a real understanding why nobody was willing to play with me. I remember I went into the room where his body lay, and my mother sat weeping alone by it. I had my battledore in my hand, and fell a-beating the coffin, and calling Papa; for, I know not how, I had some slight idea that he was locked up there. My mother catched me in her arms, and, transported beyond all patience of the silent grief she was before in, she almost smothered me in her embrace . . . She was a very beautiful woman, of a noble spirit; and there was a dignity in her grief amidst all the wildness of her transport, which, methought, struck me with an instinct of sorrow, which, before I was sensible of what it was to grieve, seized my very soul, and has made pity the weakness of my heart ever since. The mind in infancy is, methinks, like the body in embryo; and receives impressions so forcibly, that they are as hard to be removed by reason, as any mark with which a child is born, is to be taken away by future application.

RICHARD STEELE, *The Tatler*, 6 June 1710

'It is remarkable', said I, 'that, of all talents, the musical shows itself earliest; so that Mozart in his fifth, Beethoven in his eighth, and Hummel in his ninth year, astonished all near them by their performance and compositions.'

'The musical talent', said Goethe, 'may well show itself earliest of any; for music is something innate and internal, which needs little nourishment from without, and no experience drawn from life. Really, however, a phenomenon like that of Mozart remains an inexplicable prodigy.'

GOETHE to Johann Eckermann, 14 February 1831

Mrs Hannah More was fond of relating how she called at Mr Macaulay's, and was met by a fair, pretty, slight child, with abundance of light hair, about four years of age, who came to the front door to receive her, and tell her that his parents were out, but that if she would be good enough to come in he would bring her a glass of old spirits: a proposition which greatly startled the good lady, who had never aspired beyond cowslip wine. When questioned as to what he knew about old spirits, he could only say that Robinson Crusoe often had some.

G. O. Trevelyan, *The Life of* LORD MACAULAY, 1876

My first conscious memory dates from when I was four. I was being taken for a walk by the nursemaid. I was dressed in knickerbockers, with a fawn-coloured coat, and on my head was a red tam-o'-shanter —you know, the round cap with a little tail protruding from its centre, like the remains of a cut umbilical cord. And then out of the hawthorn hedge there hopped a fat toad. What a creature, with its warty skin, its big eyes bulging up, and its awkward movements! That comic toad helped to determine my career as a scientific naturalist.

JULIAN HUXLEY, *Memories*, 1970–3

It is alleged by a friend of my family that I used to suffer from insomnia at the age of four; and that when she asked me how I managed to occupy my time at night I answered: 'I lie awake and think about the past'.

RONALD KNOX, *Literary Distractions*, 1941

I did not cry, I seldom laughed and I did not make a noise; at four, I was caught putting salt in the jam; out of scientific interest rather than devilment, I suppose; anyway, it is the only crime I can remember.

JEAN-PAUL SARTRE, *Words*, 1964

At the age of four years and eight months, Louis XIV, King of France and Navarre, was not merely the master but also the owner of the goods and bodies of nineteen million men, given into his power by a decree of the almighty.

Philippe Erlanger, LOUIS XIV, 1970

5

The next stage is up to five years of age. During this period, it is not a good idea to try and teach them anything or make them do tasks that would interfere with their development.

ARISTOTLE, *Politics*, c. 345 BC

Endeavouring to make children prematurely wise is useless labour. Suppose they have more knowledge at five or six years old than other children, what use can be made of it? It will be lost before it is wanted, and the waste of so much time and labour of the teacher can never be repaid. Too much is expected from precocity, and too little performed.

SAMUEL JOHNSON, in Boswell's *Life of Johnson*, 1775

In every man there lies hidden a child between five and eight years old, the age at which naïveté comes to an end. It is this child whom one must detect in that intimidating man with his long beard, bristling eyebrows, heavy moustache, and weighty look—a captain. Even he concedes, and not at all deep down, the youngster, the booby, the little rascal, out of whom age has made this powerful monster.

PAUL VALÉRY (1871–1945)

Fascist governments have not found the five-year-old too young to regiment in uniform, to marshall in battalions preparatory to group behaviour which will be required in later years.

ARNOLD GESELL, *The First Five Years of Life*, 1940

Lords, knights and squires, the num'rous hand
    That wears the fair Miss Mary's fetters,
Were summoned, by her high command,
    To show their passion by their letters.

My pen amongst the rest I took,
    Lest those bright eyes that cannot read
Should dart their kindling fires, and look
    The power they have to be obeyed.

Nor quality, nor reputation,
    Forbid me yet my flame to tell,
Dear five years old befriends my passion,
    And I may write till she can spell.

For while she makes her silk-worms' beds
    With all the tender things I swear,
Whilst all the house my passion reads
    In papers round her baby's hair,

She may receive and own my flame,
    For though the strictest prudes should know it,
She'll pass for a most virtuous dame,
    And I for an unhappy poet.

Then too alas, when she shall tear
    The lines some younger rival sends,
She'll give me leave to write, I fear,
    And we shall still continue friends.

For as our diff'rent ages move,
   'Tis so ordained, would fate but mend it,
That I shall be past making love,
   When she begins to comprehend it.
   MATTHEW PRIOR, 'To a Child of Quality, Five Years Old, the
   Author Forty', 1704

Oh! many a time have I, a five years' Child,
A naked Boy, in one delightful Rill,
A little Mill-race severed from his stream,
Made one long bathing of a summer's day,
Basked in the sun, and plunged, and basked again
Alternate all a summer's day, or coursed
Over the sandy fields, leaping through groves
Of yellow grunsel, or when crag and hill,
The woods, and distant Skiddaw's lofty height,
Were bronzed with a deep radiance, stood alone
Beneath the sky, as if I had been born
On Indian Plains, and from my Mother's hut
Had run abroad in wantonness, to sport,
A naked Savage, in the thunder shower.
   WORDSWORTH, *The Prelude*, Book I, 1805

As I grew older I had a cart and a ball, and, when I was five or six,
two boxes of well-cut wooden bricks. With these modest but, I still
think, entirely sufficient possessions and being always summarily
whipped if I cried, did not do as I was bid or tumbled down the stairs,
I soon attained serene and secure methods of life and motion and
could pass my days contentedly in tracing the squares and comparing
the colours of my carpet.
   JOHN RUSKIN, *Praeterita*, 1885–9

My cousins the Rossettis were horrible monsters of precocity. Let
me set down here with what malignity I viewed their proficiency in
Latin and Greek at ages incredibly small. Thus, I believe, my cousin
Olive wrote a Greek play at the age of something like five.
   FORD MADOX FORD, *Ancient Lights*, 1911

Even at five years old there had been a dreadful maturity in Arthur
Bang's regard, in the deliberate way he turned his head and seemed
to reflect before he spoke.
   KINGSLEY AMIS, *One Fat Englishman*, 1963

From a very early age, perhaps the age of five or six, I knew that when I grew up I should be a writer. Between the ages of about seventeen and twenty-four I tried to abandon this idea, but I did so with the consciousness that I was outraging my true nature and that sooner or later I should have to settle down and write books.

GEORGE ORWELL, 'Why I Write', 1946

Almighty God (to Whom alone I owe the most profound gratitude) endowed me, especially in music, with such proficiency that even in my sixth year I was able to sing some masses in the choir loft, and to play a little on the harpsichord and violin.

HAYDN, 1776

I can remember, at the age of five, being told that childhood was the happiest period of life (a blank lie, in those days). I wept inconsolably, wished I were dead, and wondered how I should endure the boredom of the years to come.

BERTRAND RUSSELL, On Education, especially in Early Childhood, 1926

6

Soon after six years of age, when the child starts another phase of development (making the change from a society in embryo to a society just born), another form of existence sets in spontaneously in which the group is organized entirely on the conscious plane. Children then want to know the customs and laws which guide their conduct: they seek to have someone in control who will govern the community.

DR MARIA MONTESSORI, The Absorbent Mind, 1949

The sun and moon are small globes travelling a little way above the roofs of houses and following us about on our walks. Even the child of six to eight years does not hesitate to take this perception as truth and, curiously enough, he never thinks of asking himself whether these heavenly bodies do not follow other people.

JEAN PIAGET, The Child's Concept of Causality, 1930

I call that parent rash and wild
Who'd reason with a six-year child,
Believing little twigs are bent
By calm considered argument.

PHYLLIS McGINLEY, 'The Velvet Hand'

At six years old I remember to have read Belisarius, Robinson Crusoe and Philip Quarles—and then I found the Arabian Nights' entertainments—one tale of which (the tale of a man who was compelled to seek for a pure virgin) made so deep an impression on me that I was haunted by spectres whenever I was in the dark.

COLERIDGE, letter to Thomas Poole, 9 October 1797

I know not what I did up until five or six; I know not how I learned how to read; I only remember the first things that I read, and their effect on me. It is the period from which I date the steady development of my self-consciousness.

ROUSSEAU, Confessions, 1782

I can remember quite well how one day during dinner—I was six years old at the time—they were discussing my looks and Mamma, trying to discover something nice about my face, said that I had intelligent eyes, and a pleasant smile, and then, yielding to Papa's arguments and to the obvious, had been forced to admit that I was plain; and afterwards, when I was thanking her for the dinner, she patted my cheek and said: 'Remember, my little Nikolai, that no-one will love you for your face so you must try to be a sensible good boy.'

TOLSTOY, Childhood, Boyhood, Youth, 1852

At a much earlier period, say six or seven, I remember praying earnestly, but it was for no higher object than to be spared from the loss of a tooth.

GLADSTONE, in John Morley, The Life of Gladstone, 1903

But now I'm six, I'm clever as clever
So I think I'll be six for ever and ever.

A. A. MILNE, Now We Are Six, 1927

At the age of six or more, I had a horror of darkness that cannot be described. If I have known fear, it certainly originated there.

JULIEN GREEN, *To Leave Before Dawn*, 1969

I remember one of my first drawings, I was perhaps six, or even less. In my father's house there was a statue of Hercules with his club, and I drew Hercules. But it wasn't a child's drawing. It was a real drawing.

PICASSO, in Denis Thomas, *Picasso and His Art*, 1975

And oh! The Lad was Deathly Proud!
He never shook your Hand or Bowed,
But merely smirked and nodded thus:
How perfectly ridiculous!
Alas! That such affected Tricks
Should flourish in a Child of Six!

HILAIRE BELLOC, 'Godolphin Horne', *Cautionary Tales*, 1908

The practice of coitus was familiar to me at the age of six and seven, after which I suspended it and did not resume it till I was twenty-four; it was a common enough practice among the village children.

ERNEST JONES (biographer of Freud), *Free Associations: Memories of a Psychoanalyst*, 1959

I stopped believing in Santa Claus when I was six. Mother took me to see him in a department store and he asked for my autograph.

SHIRLEY TEMPLE

7

If I had the teaching of children up to seven years of age or thereabouts, I care not who had them afterwards.

A Jesuit divine (unattributed)

When they have passed their fifth birthday they should for the next two years learn simply by observation whatever they may be required to learn. Education after that may be divided into two stages—from the seventh year to puberty and from puberty to the completion of twenty-one years. Thus those who divide life into periods of seven years are not far wrong, and we ought to keep to the divisions that nature makes.

ARISTOTLE, *Politics, c.* 345 BC

The age of seven to eight . . . marks a considerable advance, for logical forms have entered upon the scene of the mind in perception.

JEAN PIAGET, *Judgement and Reasoning in the Child,* 1926

'Child' shall include any Boy or Girl who in the Opinion of the Justices is above the Age of Seven and under the Age of Fourteen.

Industrial School Act, 1857

> Among thise children was a wydwes sone,
> A litel clergeon, seven yeer of age,
> That day by day to scole was his wone,
> And eek also, where as he saugh th'ymage
> Of Cristes mooder, hadde he in usage,
> As hym was taught, to knele adoune and seyc
> His *Ave Marie,* as he goth by the weye.

CHAUCER, *The Prioress's Tale, c.* 1387

> Farewell, thou child of my right hand, and joy;
>   My sin was too much hope of thee, loved boy.
> Seven years thou wert lent to me, and I thee pay,
>   Exacted by thy fate, on the just day.
> O, could I lose all father now! For why
>   Will man lament the state he should envy?
> To have so soon 'scaped world's and flesh's rage,
>   And, if no other misery, yet age?
> Rest in soft peace, and, asked, say here doth lie
>   Ben Jonson his best piece of poetry;
> For whose sake, henceforth, all his vows be such,
>   As what he loves may never like too much.

BEN JONSON, 'On My First Son', 1616

> Wit's queen, (if what the poets sing be true)
> And beauty's goddess childhood never knew,
> Pallas they say sprung from the head of Jove,
> Full grown, and from the sea the Queen of Love;

But had they, Miss, your wit and beauty seen,
Venus and Pallas both had children been.
They, from the sweetness of that radiant look,
A copy of young Venus might have took:
And from those pretty things you speak have told,
How Pallas talked when she was seven years old.

POPE, 'Upon a Girl of Seven Years Old', c. 1713

So, closing the lesson book, the Mother went,
Satisfied, proud; not seeing in the intent
Blue eyes under the bold forehead her child's soul
Blaze with a loathing he could scarce control.
Day-long he sweated to obey; a clever brat,
But certain lurking habits hinted at
A rancorous hypocrisy in him;
Crossing hung corridors, musty, dim,
He'd stick his tongue out, clench against his thighs
His fists and watch light-specks under shut eyes.

RIMBAUD, 'The Seven-Year-Old Poet', 1870

'Seven years and six months!' Humpty Dumpty repeated thoughtfully. 'An uncomfortable sort of age. Now if you'd asked *my* advice, I'd have said, "Leave off at seven"—but it's too late now.'

'I never ask advice about growing,' Alice said indignantly.

'Too proud?' the other enquired.

Alice felt even more indignant at this suggestion. 'I mean,' she said, 'that one can't help growing older.'

'*One* can't, perhaps,' said Humpty Dumpty, 'but *two* can. With proper assistance, you might have left off at seven.'

LEWIS CARROLL, Through the Looking Glass, 1872

At the age of seven years, the Spartan boy was consigned to the care of a state-officer, and the course of his education was entirely determined by the purpose of inuring him to bear hardships, training him to endure an exacting discipline, and instilling into his heart a devotion to the state. The boys, up to the age of twenty, were marshalled in a huge school formed in the model of an army.

J. B. BURY, A History of Greece, 1900

I have been this morning with Lady Hester Pitt and there is little William Pitt, not eight years old and really the cleverest child I ever saw, and brought up so strictly, and so proper in his behaviour that, mark my words, that little boy will be a thorn in Charles's side as long as he lives.

WILLIAM PITT, described in a letter from Lady Holland, 1766

8

This is the last letter I shall write to you as a little boy: for, tomorrow, if I am not mistaken, you will attain your ninth year; so that for the future, I shall treat you as a *youth*. You must now commence a different course of life, a different course of studies. No more levity: childish toys and playthings must be thrown aside, and your mind directed to serious objects. What was not unbecoming of a child would be disgraceful to a youth.

LORD CHESTERFIELD, letter to his son Philip Stanhope (translated from the Latin), May 1741

Not to expose your true feelings to an adult seems to be instinctive from the age of seven or eight onwards. Even the affection that one feels for a child, the desire to protect and cherish it, is a cause of misunderstanding. One can love a child, perhaps, more deeply than one can love another adult, but it is rash to assume that the child feels any love in return. Looking back on my own childhood, after the infant years were over, I do not believe that I ever felt love for any mature person, except my mother, and even her I did not trust, in the sense that shyness made me conceal most of my real feelings from her.

GEORGE ORWELL, 'Such, Such Were the Joys', 1947

You should not take a fellow eight years old
And make him swear to never kiss the girls.

ROBERT BROWNING, 'Fra Lippo Lippi', c. 1850

Under seven years of age, indeed, an infant cannot be guilty of a felony, for then a felonious indiscretion is almost an impossibility in nature; but at eight years old he may be guilty of a felony.

SIR WILLIAM BLACKSTONE, *Commentaries*, 1769

In my ninth year (January 1746) . . . I was sent to Kingston-upon-Thames to a school of about seventy boys, which was kept by Dr Woodeson and his assistants. Every time I have since passed over Putney Common, I have always noticed the spot where my mother, as we drove along in the coach, admonished me that I was now going into the world and must learn to think and act for myself.

EDWARD GIBBON, *Autobiography*, 1764–91

I remember that at eight years old I walked with him [his father] one winter evening from a farmer's house a mile from Ottery . . . and he told me the names of the stars and how Jupiter was a thousand times larger than our world, and that the other twinkling stars were suns that had worlds rolling round them; and when I came home he showed me how they rolled round. I heard him with a profound delight and admiration; but without the least mixture of wonder or incredulity. For from my early reading of fairy tales and genii, etc., etc., my mind had been habituated *to the vast*, and I never regarded *my senses* in any way as the criteria of my belief. I regulated all my creeds by my conceptions, not by my *sight*, even at that age.

COLERIDGE, letter to Thomas Poole, 16 October 1797

(In my eighth year) I had read, under my father's tuition, a number of Greek prose authors, among whom I remember the whole of Herodotus, and of Xenophon's Cyropaedia and Memorials of Socrates; some of the lives of the philosophers by Diogenes Laertius; part of Lucian, and Isocrates' ad Demonicum and ad Nicoclem.

JOHN STUART MILL, *Autobiography*, 1873

'But even before I learned to read, I remember how I was first moved by deep spiritual emotion when I was eight years old. My mother took me alone to church (I don't remember where my brother was at the time), to morning mass on the Monday before Easter. It was a sunny day, and I remember now, just as though I saw it again, how the incense rose from the censer and floated slowly upwards and how through a little window from the dome above the sunlight streamed down upon us and, rising in waves towards it, the incense seemed to dissolve in it. I looked, and felt deeply moved and for the first time in my life I consciously received the first seed of the words of God in my soul.'

Father Zossima in DOSTOYEVSKY's *The Brothers Karamazov*, 1880

Somehow, mysteriously, when I was eight years old or so, the soldiery was eclipsed for me by the constabulary. Somehow the scarlet and the bearskins began to thrill me less than the austere costume and calling of the Metropolitan Police. Once in every two hours a policeman came, on his beat, past the house of my parents. At the window of the dining-room I would await his coming punctually, behold him with profound interest, and watch him out of sight. It was not the daffodils that marked for me the coming of the season of Spring. It was the fact that policemen were wearing long thick frock coats with buttons of copper . . . But the young are faithless. By the time I was eleven years old I despised the Force. I was interested only in politicians—in Statesmen, as they were called in that time.

MAX BEERBOHM, 'A Small Boy Seeing Giants', broadcast 26 July 1936

'I'm a trapper in the Gamber Pit, I have to trap without a light and I'm scared. I go at four and sometimes half-past-three in the morning and come out at five and half past. I never go to sleep. Sometimes I sing when I've light, but not in the dark: I dare not sing then.'

A child of eight quoted in the *First Report on the Employment of Children and Young Persons in Mines*, 1842

After the age of eight, there could be no more school, for now all the older children had gone out into the world to make their own poor living, the boys to work on distant farms, the girls to service or to be wives, and Joan was wanted at home to keep house for her father, to do the washing, mending, cleaning, cooking, and to be mother to her little brother as well.

W. H. HUDSON, *A Shepherd's Life*, 1910

Nine times already since my birth had the heaven of light returned to the selfsame point almost, as concerns its own revolution, when first the glorious Lady of my mind was made manifest to mine eyes; even she who was called Beatrice by many who knew not wherefore. She had already been in this life for so long as that, within her time, the starry heaven had moved towards the Eastern quarter one of the twelve parts of a degree; so that she appeared to me at the beginning of her ninth year almost, and I saw her almost at the end of my ninth year. Her dress, on that day, was of a most noble colour, a subdued and goodly crimson, girdled and adorned in such sort as best suited with her very tender age. At that moment, I say most truly that the spirit of life, which hath its dwelling in the secretest chamber of the heart, began to tremble so violently that the least pulses of my body shook therewith; and in trembling it said these words: 'Here is a deity stronger than I; who, coming, shall rule over me'.

DANTE, *Vita nuova*, 1290–4, trans. Dante Gabriel Rossetti

9

He [John Donne] had his first breeding in his father's house, where a private tutor had the care of him until the ninth year of his age; and, in his tenth year, was sent to the University of Oxford, having at that time a good command both of the French and Latin tongue. This, and some other of his remarkable abilities, made one give this censure of him, *That this age had brought forth another Picus Mirandula*; of whom story says, *That he was rather born than made wise by study*.

IZAAK WALTON, *The Life of Donne*, 1640

Oliver Twist's ninth birthday found him a pale, thin child, some-what diminutive in stature, and decidedly small in circumference. But nature or inheritance had implanted a good sturdy spirit in Oliver's breast. It had had plenty of room to expand, thanks to the spare diet of the establishment; and perhaps to this circumstance may be attributed his having any ninth birthday at all. Be this as it may, however; it *was* his ninth birthday; and he was keeping it in the coal-cellar with a select party of two other young gentlemen, who after participating with him in a sound thrashing, had been locked up for atrociously presuming to be hungry.

CHARLES DICKENS, *Oliver Twist*, 1839

The attic was a favourite retreat on a wet day . . . here she fretted out all her ill-humours . . . and here she kept a fetish which she punished for all her misfortunes. This was the trunk of a large wooden doll which once stared with the roundest of eyes above the reddest of cheeks but was now entirely defaced by a long career of vicarious suffering. Three nails driven into the head commemorated as many crises in Maggie's nine years of earthly struggle; that luxury of vengeance having been suggested to her by the picture of Jael destroying Sisera in the old Bible.

GEORGE ELIOT, *The Mill on the Floss*, 1860

'Your brother is not interested in ancient monuments?' Winter-
bourne inquired, smiling.
    'He says he don't care much about old castles. He's only nine.'
HENRY JAMES, Daisy Miller, 1879

It was in 1868, when nine years old or thereabouts, that while
looking at a map of Africa of the time and putting my finger on the
blank space then representing the unsolved mystery of that conti-
nent, I said to myself with absolute assurance and an amazing
audacity which are no longer in my character now: 'When I grow up,
I shall go there.'
JOSEPH CONRAD, A Personal Record, 1912

Between the age limits of nine and fourteen there occur maidens
who, to certain bewitched travellers, twice or many times older than
they, reveal their true nature, which is not human but nymphic (that
is demoniac); and these chosen creatures I propose to designate as
'nymphets'.
VLADIMIR NABOKOV, Lolita, 1955

When I was nine I played the demon king in Cinderella and it
launched me on a long and happy life of being a monster.
BORIS KARLOFF

10

The ten-year-old gives a fair indication of the man or woman he or she is to be.

ARNOLD GESELL, *The Child from Five to Ten*, 1946

At ten, Mercury is in the ascendant; and at that age, a man, like this planet, is characterized by extreme mobility within a narrow sphere, where trifles have a great effect upon him.

SCHOPENHAUER, 'Counsels and Maxims', *Parerga and Paralipomena*, 1851

In about their tenth or eleventh year, children get to hear about sexual matters. A child who has grown up in a comparatively uninhibited social atmosphere, or who has found better opportunities for observation, tells other children what he knows, because this makes him feel mature and superior.

SIGMUND FREUD, *On the Sexual Theories of Children*, 1908

At ten years old I was taken home to assist my father in his business, which was that of a tallow chandler and soap-boiler.

BENJAMIN FRANKLIN, *Autobiography*, 1771–90

Ah! when I was ten years old I beat you all—Napoleon and all—in ambition!

ELIZABETH BARRETT BROWNING, letter to R. H. Horne

I have now reached my eleventh year, and since then my Muse in hours of sacred inspiration has often whispered to me: 'Make the attempt, just put down on paper the harmonies of your soul!' —Eleven years—I thought—and how could I look like a composer? And what would the experienced adults in this art say to this? I was almost too shy. But my Muse insisted—I obeyed and I composed.

BEETHOVEN, dedication letter to the Prince Elector Max Friedrich of Cologne, 1783

But soon as Luke, full ten years old, could stand,
Against the mountain blasts; and to the heights,
Not fearing toil, nor length of weary ways,
He with his Father daily went, and they
Were as companions, why should I relate
That objects which the Shepherd loved before
Were dearer now? that from the Boy there came
Feelings and emanations—things which were
Light to the sun and music to the wind;
And that the old Man's heart seemed born again?

WORDSWORTH, 'Michael', 1800

This little maid had just entered her eleventh year; but her important station at the theatre, as it seemed to her, with the benefits which she felt to accrue from her pious allocation of her small earnings, had given an air of womanhood to her steps and to her behaviour. You would have taken her to be at least five years older.

Till lately, she had merely been employed in choruses, or where children were wanted to fill up the scene. But the manager, observing a diligence and adroitness in her above her age, had for some months past intrusted to her the performance of whole parts. You may guess the self-consequence of the promoted Barbara . . . At the period I commenced with, her slender earnings were the sole support of the family, including two younger sisters.

CHARLES LAMB, 'Barbara S————', *The Last Essays of Elia*, 1833

'She is ten years of age, sir.'
'Not more!'
'Not a day.'
'Dear me!' said Nicholas, 'it's extraordinary!'

It was; for the infant phenomenon, though of short stature, had a comparatively aged countenance, and had moreover been precisely the same age—not perhaps to the full extent of the memory of the oldest inhabitant, but certainly for five good years. But she had been kept up late every night, and put upon an unlimited allowance of gin-and-water from infancy to prevent her growing tall.

CHARLES DICKENS, *Nicholas Nickleby*, 1839

Penrod approached his twelfth year wearing an expression carefully trained to be inscrutable. Since the world was sure to misunderstand everything, mere defensive instinct prompted him to give it as little

as possible to lay hold upon. Nothing is more impenetrable than the
face of a boy who has learned this.

BOOTH TARKINGTON, *Penrod*, 1914

Why, there was idle Jerry Jones,
As dirty as a pig,
Who smoked when only ten years old,
And thought it made him big.

He'd puff along the open street,
As if he had no shame;
He'd sit beside the tavern door,
And there he'd do the same.

He spent his time and money too,
And made his mother sad;
She feared a worthless man would come
From such a worthless lad.

from *The Temperance Orator and Reciter*, 19th Century

11

There is a tide which begins to rise in the veins of youth at the age of
eleven or twelve. It is called by the name of adolescence. If that tide
can be taken at the flood, and a new voyage begun in the strength and
along the flow of its current, we think that it will move on to fortune.
We therefore propose that all children should be transferred, at the
age of eleven or twelve, from the junior or primary school.

The Hadow Report, justifying the eleven-plus examination,
1926

In February I'll be twelve. How old I am!

ANAÏS NIN, *Diaries*, 2 January 1915

He was almost twelve. He was done with childhood. As that spring
ripened he felt entirely, for the first time, the full delight of loneliness. Sheeted in the thin nightgown, he stood in darkness by the
orchard window . . . drinking the sweet air down, exulting in his

isolation in darkness, hearing the strange wail of the whistle going west.

THOMAS WOLFE, *Look Homeward, Angel*, 1929

> When eleven or twelve,
> one year short
> of the catastrophic brink of adolescence,
> I nightly enjoyed my mother bathing—
> not lust, but lust of the eye.

ROBERT LOWELL, 'Day by Day', *The Art of the Possible*, 1978

My father destined me, while yet a little child, for the study of humane letters, which I seized with such eagerness that from the twelfth year of my age I scarcely ever went from my lessons to bed before midnight; which, indeed, was the first cause of injury to my eyes, to whose natural weakness there were also added frequent headaches.

MILTON, *Defensio Secunda*, 1654

The first sense he had of God was when he was eleven years old at Chigwell, being retired in a chamber alone; he was so suddenly surprised with an inward comfort and (as he thought) an external glory in the room that he has many times said that from thence he had the seal of divinity and immortality, that there was a God and that the soul of man was capable of enjoying his divine communications.

WILLIAM PENN (1644–1718) in John Aubrey, *Brief Lives*

For I remember when I began to read and to take some pleasure in it, there was wont to lie in my mother's parlour (I know not by what accident, for she herself never in her life read any book but of devotion) but there was wont to lie Spenser's Works. This I happened to fall upon, and was infinitely delighted with the stories of the knights, and giants, and monsters, and brave houses, which I found everywhere there: (tho' my understanding had little to do with all this) and by degrees with the tinkling of the rhyme and the dance of the numbers, so I think I had read him all over before I was twelve years old, and was thus made a Poet as irremediably as a child is made an eunuch.

ABRAHAM COWLEY, *Of Myself*, 1668

Somewhere along the line, the child must become possessed by music, by the sudden desire to play, to excel. It can happen at any time between the ages of ten or so and fourteen. Suddenly the child begins to sense something happening and he really begins to work, and in retrospect the first five or six years seem like kinderspiel, fooling around. At this point the prodigy begins to flower. It happened to me when I was eleven.

ISAAC STERN, violinist, *New York Times*, January 1980

It was midway through his schooldays that there came to Napoleon what often comes to boys of that age (his twelfth to thirteenth year)—at least to boys of the Catholic culture and surrounded by an unescapable routine of religious practice—the failure of religious faith.

HILAIRE BELLOC, *Napoleon*, 1932

At the age of eleven, I began Euclid, with my brother as my tutor. This was one of the great events of my life, as dazzling as first love. I had not imagined that there was anything so delicious in the world. After I had learned the fifth proposition, my brother told me that it was generally considered difficult, but I had found no difficulty whatever. This was the first time it had dawned upon me that I might have some intelligence.

BERTRAND RUSSELL, *Autobiography*, i, 1967

For the first time in my life—I was then eleven years old—I felt myself forced into open opposition. No matter how hard and determined my father might be about putting his own plans and opinions into action, his son was no less obstinate in refusing to accept ideas on which he set little or no value.

I would not become a civil servant.

ADOLF HITLER, *Mein Kampf*, 1933

12

All is accomplished by the time we are twelve years old.

CHARLES PÉGUY (1873–1914)

Give me a child of twelve who knows nothing at all, and at fifteen I will give him back to you as wise as the one you have instructed from the beginning.

ROUSSEAU, *Émile*, 1762

I remember myself, at twelve years old, as a complete child, in spite of my reading of novels; a year later I already liked working; and the capacity for thought had already awoken in my soul, which till then had lived only by a child's imagination.

ALEXANDER HERZEN, *My Past and Thoughts*, 1852–3

In 1830 the average age at menarche was seventeen: today it is between twelve and thirteen.

PAULA WEIDEGER, *Female Cycles*, 1978

> 'Experience, though noon auctoritee
> Were in this world, is right ynogh for me
> To speke of wo that is in mariage;
> For, lordynges, sith I twelve yeer was of age,
> Thonked be God that is eterne on lyve,
> Housbondes at chirche dore I have had fyve,—
> If I so ofte myghte have ywedded bee,—
> And alle were worthy men in hir degree.'

CHAUCER, *The Wife of Bath's Prologue*, c. 1387

I had the pleasure, on this occasion, of renewing the acquaintance of Master Micawber, whom I found a promising boy of about twelve or thirteen, very subject to that restlessness of limb which is not an unfrequent phenomenon in youths of his age.

CHARLES DICKENS, *David Copperfield*, 1849–50

On a search night, when the constables have taken up near forty prostitutes, it has appeared on their examination, that the major part of them have been of this kind, under the age of eighteen, many not more than twelve, and these, though young, half eaten up with the foul distemper.

SIR JOHN FIELDING, *An Account of the Origin and Effects of a Police Force*, 1758

I have read somewhere that children from twelve to fourteen years of age—that is, in the transition stage from childhood to adolescence —are singularly inclined to arson and even murder. As I look back upon my boyhood . . . I can quite appreciate the possibility of the most frightful crime being committed without object or intent to injure but *just because*—out of curiosity, or to satisfy an unconscious craving for action.

TOLSTOY, *Childhood, Boyhood, Youth*, 1852

He will not endure (albeit he does not confess so much) to be told to do anything, at least in that citadel of freedom, his home. His elders probably give him as few orders as possible. He will almost ingeniously evade any that are inevitably or thoughtlessly inflicted upon him, but if he does but succeed in only postponing his obedience, he has, visibly, done something for his own relief. It is less convenient that he should hold mere questions, addressed to him in all good faith, as in some sort an attempt upon his liberty.

ALICE MEYNELL, 'The Boy', *The Children*, 1897

I had scarcely passed my twelfth birthday when I entered the inhospitable regions of examinations, through which for the next seven years I was destined to journey. These examinations were a great trial to me. The subjects which were dearest to the examiners were almost invariably those I fancied least. I would have liked to have been examined in history, poetry and writing essays. The examiners, on the other hand, were partial to Latin and mathematics. But their will prevailed . . .

SIR WINSTON CHURCHILL, *My Early Life*, 1930

Of Sarah Byng the tale is told
How when the child was twelve years old
She could not read or write a line.
Her sister Jane, though barely nine,
Could spout the Catechism through
And parts of Matthew Arnold too,

While little Bill who came between
Was quite unnaturally keen
On 'Athalie', by Jean Racine.
But not so Sarah! Not so Sal!
She was a most uncultured girl
Who didn't care a pinch of snuff
For any literary stuff
And gave the classics all a miss . . .

HILAIRE BELLOC, *New Cautionary Tales*, 1931

Sad and terrible happenings had never made Frankie cry, but this season many things made Frankie suddenly wish to cry. Very early in the morning she would sometimes go out into the yard and stand for a long time looking at the sunrise sky. And it was as though a question came into her heart, and the sky did not answer. Things she had never noticed much before began to hurt her; home lights watched from the evening sidewalks, an unknown voice from an alley. She would stare at the lights and listen to the voice, and something inside her stiffened and waited. But the lights would darken, the voice fall silent, and though she waited that was all. She was afraid of these things that made her suddenly wonder who she was, and what she was going to be in the world, and why she was standing at that minute seeing a light, or listening, or staring up into the sky; alone. She was afraid, and there was a queer tightness in her chest.

CARSON McCULLERS, *The Member of the Wedding*, 1952

He was old enough, twelve years and a few months, to have lost the prominent tummy of childhood; and not yet old enough for adolescence to have made him awkward.

WILLIAM GOLDING, *The Lord of the Flies*, 1954

Through the reading of popular scientific books I soon reached the conviction that much of the stories in the Bible could not be true. The consequence was a positively fanatic orgy of free-thinking coupled with the impression that youth is intentionally being deceived by the state through lies; it was a crushing impression. Suspicion against every kind of authority grew out of this experience, a sceptical attitude towards the convictions which were alive in any specified social environment—an attitude which has never again left me.

ALBERT EINSTEIN

13

From the age of thirteen I had revelations from our Lord by a voice
which told me how to behave.

> Minute of the Interrogation of JOAN OF ARC, 22 February 1451

> Well, think of marriage now; younger than you,
> Here in Verona, ladies of esteem,
> Are made already mothers: by my count
> I was your mother much upon these years
> That you are now a maid. Thus then in brief:
> The valiant Paris seeks you for his love.
>> Lady Capulet to her daughter Juliet, who was to be fourteen in
>> 'a fortnight and odd days'. SHAKESPEARE, *Romeo and Juliet*,
>> 1597

> Thirteen years
> Or haply less, I might have seen, when first
> My ears began to open to the charm
> Of words in tuneful order, found them sweet
> For their own sakes, a passion and a power;
> And phrases pleased me, chosen for delight,
> For pomp, or love.
>> WORDSWORTH, *The Prelude*, Book V, 1805

I remember well the spot where I read these volumes [Bishop Percy's
Reliques]. It was beneath a huge platanus-tree in the ruins of what had
been intended for an old-fashioned arbour . . . The summer day
sped onward so fast, that notwithstanding the sharp appetite of
thirteen, I forgot the hour of dinner, was sought for with anxiety,
and was still found entranced in my intellectual banquet . . . To this
period also I can trace distinctly the awaking of the delightful feeling
for the beauties of natural objects which has never since deserted me.

> SIR WALTER SCOTT, 'A Memoir', from J. G. Lockhart's *Life of
> Scott*, 1837–8

On the 16th February, 1723, the King having entered upon his
fourteenth year, the Duc d'Orléans attended his levee in order to pay
his respects and to ask his commands for the conduct of State affairs.

The ceremony was followed by another of a more impressive character. This was the Bed of Justice held on the 22nd February following, at which his Majesty declared his majority, and stated that, in accordance with the regulations of the State, he had come to his Parliament to announce to it that from that day forth it was his intention to assume the direction of the Government.

> LOUIS XV, from Mouffle D'Angerville, *The Private Life of Louis XV*, 1921

It is my painful duty to have to record here my marriage at the age of thirteen. As I see the youngsters of the same age about me who are under my care, and think of my own marriage, I am inclined to pity myself and congratulate them on having escaped my lot. I can see no moral argument in support of such a preposterously early marriage.

> GANDHI, *An Autobiography*, 1927

I had passed thirteen, and things were worse even than I had foreseen. I lay in bed in the dormitory of St John's, listening to the footsteps clatter down the stone stairs to early prep. and breakfast, and when the silence had safely returned I began trying to cut my right leg open with a penknife. But the knife was blunt and my nerve was too weak for the work.

> GRAHAM GREENE, *A Sort of Life*, 1971

In a general way the whole period of education is dominated by this threefold rhythm. Till the age of thirteen or fourteen there is the romantic stage, from fourteen to eighteen the stage of precision, and from eighteen to two and twenty the stage of generalization.

> A. N. WHITEHEAD, *The Aims of Education and Other Essays*, 1929

The second stage of initiation in the Jewish tradition is the Bar Mitzvah ceremony. Traditionally this takes place on the sabbath nearest to a boy's thirteenth birthday. In its present form the service has been used since about the fourteenth century.

The term 'Bar Mitzvah' actually applies to every adult Jew, and its literal meaning is 'son of the commandment'. It is now celebrated by formally calling the boy to the Torah scroll during the sabbath service, where he recites the blessing before and after the reading, and usually himself reads a portion in Hebrew. There is also a custom

that when the boy has completed the reading his father says: 'Blessed
is he who has absolved me from responsibility for this child.'

John Prickett, *Initiation Rites*, 1978

Oh the innocent girl
in her maiden teens
knows perfectly well
what everything means.

If she didn't, she oughter;
it's a silly shame
to pretend that your daughter
is a blank at the game.

D. H. LAWRENCE, 'The Jeune Fille', *Pansies*, 1929

Thirteen's anomalous—not that, not this:
Not folded bud, or wave that laps a shore,
Or moth proverbial from the chrysalis.
Is the one age defeats the metaphor.
Is not a town, like childhood, strongly walled
But easily surrounded; is no city.
Nor, quitted once, can it be quite recalled—
Not even with pity.

PHYLLIS McGINLEY, 'A Certain Age'

Until the rise of American advertising, it never occurred to anyone
anywhere in the world that the teenager was a captive in a hostile
world of adults, with his own morality and language, his difference
from the world of adults a source of pride.

GORE VIDAL, *Rocking the Boat*, 1962

14

At fourteen he wedded was,
A father at fifteen,
At sixteen's face was white as milk,
And then his grave was green;
And the daisies were outspread,

And buttercups of gold,
O'er my pretty lad so young
Now ceased growing.

                                                              Anon.

Neither must we think, that the life of Man begins when he can feed
himself or walk alone, when he can fight, or beget his like; for so he is
contemporary with a camel, or a cow; but he is first a man when he
comes to a certain, steady use of reason, according to his proportion,
and when that is, all the world of men cannot tell precisely. Some are
called *at age*, at fourteen, some at one and twenty, some never; but all
men, late enough; for the life of a man comes upon him slowly and
insensibly.

          JEREMY TAYLOR, *The Rule and Exercises of Holy Dying*, 1650

Depend upon it, a man never experiences such pleasure or grief after
fourteen years as he does before, unless in some cases, in his first
love-making, when the sensation is new to him.

                                        CHARLES KINGSLEY, 1819–75

When she reached her fifteenth year her beauty began to shine forth
like the noonday sun; which it surpassed in brightness. As for her
mind, it was equal in beauty to her body: for she was well-versed in
Latin, and when she was between thirteen and fourteen years old she
delivered a public oration in Latin before the King and Queen and the
whole Court in the great hall of the Louvre, maintaining and
upholding against the common opinion that it was becoming in
women to be skilful in the liberal arts and sciences.

     MARY QUEEN OF SCOTS, in Pierre de Brantôme, *Memoirs*, 1665–6

Michelangelo was at this time fourteen years old, and he made such
progress that he astonished Domenico [Ghirlandaio], who saw that
he not only surpassed his other pupils, of whom he had a great
number, but often equalled the things he did himself.

               GIORGIO VASARI, *Lives of the Painters*, 1550, 1568

I hid myself within myself, I only considered myself and quietly
wrote down all my joys, sorrows and contempt in my diary . . . I
used to be furious with Mummy, and still am sometimes. It's true
that she doesn't understand me, but I don't understand her either.

                          ANNE FRANK, *Diary*, January, 1944

At fourteen I fell into one of those fits of bottomless despair that come with adolescence, and I seriously thought of dying because of the mediocrity of my natural faculties.

SIMONE WEIL, *Gravity and Grace*, 1948

Fourteen-year-old, why must you giggle and dote,
Fourteen-year-old, why are you such a goat?
I'm fourteen years old, that is the reason,
I giggle and dote in season.

STEVIE SMITH, 'The Conventionalist'

Last year, when Jeffrey turned fourteen and Matilda twelve, they had begun to change; to grow rude, coarse, selfish, insolent, nasty, brutish and tall. It was as if she were keeping a boarding house in a bad dream, and the children she loved were turning into awful lodgers—lodgers who paid no rent, whose leases could not be terminated. They were awful at home and abroad; in company and alone; in the morning, the afternoon and the evening.

ALISON LURIE, *The War Between the Tates*, 1974

I was a fourteen-year-old boy for thirty years.

MICKEY ROONEY

15

Sylvia the fair, in the bloom of Fifteen,
Felt an innocent warmth, as she lay on the green;
She had heard of a pleasure, and something she guest
By the towzing and tumbling and touching her Breast;
She saw the men eager, but was at a loss,
What they meant by their sighing and kissing so close;
By their praying and whining
And clasping and twining,
And panting and wishing,
And sighing and kissing
And sighing and kissing so close.

JOHN DRYDEN, 'A New Song', 1685

Common beauties stay fifteen.

ANDREW MARVELL, 'Young Love', 1681

'A woman by the time she is fifteen should know all about managing men.'

Despina in MOZART's *Così Fan Tutte*, 1790

'Sono quindici!'

Madame Butterfly in PUCCINI's opera, 1904

At fifteen appearances were mending; she began to curl her hair and long for balls; her complexion improved, her features were softened by plumpness and colour, her eyes gained more animation, and her figure more consequence. Her love of dirt gave way to an inclination for finery, and as she grew clean she grew smart.

JANE AUSTEN, *Northanger Abbey*, 1818

Most people, I suppose, would, like the honourable proposer, fix the prime of a man's life somewhere about thirty or thirty-five. Personally—I am open to conversion and do not hold this out as an essential article in my creed—I should place it at between fifteen and sixteen. It is then, it always seems to me, that his vitality is at its highest; he has greatest sense of the ludicrous and least sense of dignity. After that time, decay begins to set in.

EVELYN WAUGH, at the age of sixteen, in a debate at Lancing School, September 1920

I shall not say why and how I became, at the age of fifteen, the mistress of the Earl of Craven. Whether it was love, or the severity of my father, or the winning arts of the noble lord, which induced me to leave my parental roof and place myself under his protection, does not now much signify; or, if it does, I am not in the humour to gratify curiosity in the matter.

HARRIETTE WILSON, courtesan, *Memoirs*, 1825

She was an indefatigable student; constantly reading and learning; with a strong conviction of the necessity and value of education, very unusual in a girl of fifteen.

Elizabeth Gaskell, *Life of* CHARLOTTE BRONTË, 1857

When I was fifteen (in the autumn of 1816), a great change of thought took place in me. I fell under the influences of a definite Creed, and received into my intellect impressions of dogma, which through God's mercy, have never been effaced or obscured.

J. H. NEWMAN, *Apologia pro Vita Sua*, 1864

I knew at fifteen pretty much what I wanted to do . . . I resolved that at thirty I would know more about poetry than any man living.

EZRA POUND, *How I Began*, 1913

> Kicking a little stone, he turned to me
> And said, 'Tell me, do you write poetry?'
> I never had, and said so, but I knew
> That very moment what I wished to do.

W. H. AUDEN, 'Letter to Lord Byron', Part IV, 1936

Fifteen-year-old Jo was very tall, thin, and brown, and reminded one of a colt; for she never seemed to know what to do with her long limbs, which were very much in her way. She had a decided mouth, a comical nose, and sharp, grey eyes, which appeared to see everything, and were by turns fierce, funny or thoughtful. Her long, thick hair was her one beauty; but it was usually bundled into a net, to be out of the way. Round shoulders had Jo, big hands and feet, a fly-away look to her clothes, and the uncomfortable appearance of a girl who was rapidly shooting up into a woman, and didn't like it.

LOUISA M. ALCOTT, *Little Women*, 1868

From the time I knew I was mortal I found the idea of death terrifying. Even when the world was at peace and my happiness seemed secure, my fifteen-year-old self would often turn at the thought of that utter non-being—*my* utter non-being—that would descend on its appointed day, for ever and ever. This annihilation filled me with such horror that I could not conceive the possibility of facing it coolly. What people called 'courage' I could only regard as blind frivolousness.

SIMONE DE BEAUVOIR, *The Prime of Life*, 1960

In a discreet and anonymous enquiry into the daydreams of a class of fifteen-year-olds, a French teacher found future marshals of France or presidents of the Republic, great men of all kinds, among the most

timid and serious boys, some of whom already saw their statues in
the squares of Paris.

JEAN PIAGET, *Six Psychological Studies*, 1968

Everything you are and do from fifteen to eighteen is what you are
and will do through life.

SCOTT FITZGERALD, letter to his daughter, 19 September 1938

16

Age? Sixteen. The very flower of youth.

TERENCE, *Eunuchus*, 161 BC

That boiling, boisterous part of life . . .

JOHN LOCKE, *Some Thoughts on Education*, 1693

At sixteen, the adolescent knows about suffering because he himself
has suffered, but he barely knows that other beings also suffer; seeing
without feeling is not knowledge.

ROUSSEAU, *Émile*, 1762

Every moment that you now lose, is so much character and advantage lost; as, on the other hand, every moment that you now employ
usefully, is so much time wisely laid out, at most prodigious interest.
These two years must lay the foundations of all the knowledge that
you will ever have; you may build upon them afterwards as much as
you please, but it will be too late to lay any new ones.

LORD CHESTERFIELD, letter to his son Philip Stanhope, May 1748

For me the time was approaching when childhood comes to an end
and youth begins: this usually happens at sixteen. The *naïve* charm of
the child vanishes and that of the young man does not yet appear.
There is a disharmony in the features and they become coarser; there
is no grace; the voice modulates from thin to thick; the eyes are
languid, but now and again they flame: physical maturity is drawing
near. The same happens in the spirit: undefined emotions, embryos
of passions, agitation, languor, a sensation of something secret and

mysterious and, following upon this, youth, lyrical enthusiasm full of love, embraces open to welcome the whole of God's world.

ALEXANDER HERZEN, *Notes of a Young Man*, 1840

Fists probing my torn pockets, off I'd wander,
My overcoat more holes than cloth and I
Poetry's bondslave under the open sky;
Oh lord, what splendid dreams of love I'd squander!
Torn were my only trousers at the knees;
As inn-sign in the sky, the Great Bear shone
And stars, like twinkling silk, smiled down upon
Tom Thumb, the dreamer, shelling rhymes like peas.

RIMBAUD, 'Ma Bohème', 1870

Stephen watched the three glasses being raised from the counter as his father and his two cronies drank to the memory of their past. An abyss of fortune or of temperament sundered him from them. His mind seemed older than theirs: it shone coldly on their strifes and happiness and regrets like a moon upon a younger earth. No life or youth stirred in him as it had stirred in them. He had known neither the pleasures of companionship with others nor the vigour of rude male health nor filial piety. Nothing stirred within his soul but a cold and cruel and loveless lust. His childhood was dead or lost and with it his soul capable of simple joys and he was drifting amid life like the barren shell of the moon.

JAMES JOYCE, *A Portrait of the Artist as a Young Man*, 1916

I laugh at you when you say, 'What if Anthony were sixteen, and read this novel?' [*Lady Chatterley's Lover*]. He'd be too bored at sixteen: but at twenty, of course, he *should* read it. Was your mind a sexual blank at sixteen? Is anybody's? And what ails the mind in that respect is that it has nothing to go on, it grinds away in abstraction.

D. H. LAWRENCE, letter to Juliette Huxley, 1928

His father supposed him a prey to the first flutterings of sex. At sixteen I was the same, he would say. At sixteen you were earning your living, said his wife. So I was, said Mr Saposcat.

SAMUEL BECKETT, *Malone Dies*, 1951

Up to sixteen years the tendency to suicide is very slight, due to age, without considering other factors.

ÉMILE DURKHEIM, *Suicide: A Study in Sociology*, 1897

The best thing to do is to behave in a manner befitting one's age. If you are sixteen or under, try not to go bald.

WOODY ALLEN, *Without Feathers*, 1976

17

From seventeen to seven and twenty (the most dangerous time of all a man's life).

ROGER ASCHAM, *The Scholemaster*, 1570

Accordingly we conclude that the appropriate age for marriage is about the eighteenth year for girls and for men the thirty-seventh plus or minus.

ARISTOTLE, *Politics*, c. 345 BC

I was not yet in love, but I was in love with love, and, from the depths of my need, I hated myself for not more keenly feeling the need . . .

ST AUGUSTINE, *Confessions*, 397–401

Experience! What, at seventeen?

Sir Lucius O'Trigger in SHERIDAN's *The Rivals*, 1775

How could I help her? 'Would I—was it wrong?'
(Claspt hands and that petitionary grace
Of sweet seventeen subdued me as she spoke) . . .

TENNYSON, 'The Brook', 1864

I am one of those people always disgusted from one day to the next, always thinking of the future, always dreaming, or rather day-dreaming, surly, pestiferous, never knowing what they want, bored with themselves and boring to everybody else. I went to the brothel for some fun, and was merely bored. Magnier [his Professor of

Rhetoric] gets on my nerves, history I find oppressive. Tobacco? My throat is raw from it. Alcohol? I am pickled in it. The only thing left is eating.

FLAUBERT, letter to Ernest Chevalier, 1839

And so it came about that at seventeen I wilfully withdrew from the society of those about me. The laws of property and inheritance, murderous oppression, the provocation of wars; the privileges of fortune and education; the prejudices of rank, and those of moral intolerance; the childish idleness of people of fashion, the brutishness of avarice; whatever remains of pagan institutions or customs in a self-styled Christian society—all those revolted me so deeply that my soul was prompted to protest against the work of the centuries. I hadn't the notion 'progress'—it wasn't popular then, and it hadn't reached me through my reading—so I saw no way out of my anguish, and the idea of working, even in my obscure and closely bounded social environment, to redeem the promises of the future could scarcely occur to me.

GEORGE SAND, *My Life*, 1854

'Dear me', Mrs Carthew went on, 'I'm glad I'm not going to school for the first time; you won't like it at all at first, and then you'll like it very much indeed, and then you'll go on liking it very much or you'll hate it. If you go on liking it—I mean when you're quite old —sixteen or seventeen—you'll never do anything, but if you hate it then, you'll have a chance of doing something.'

COMPTON MACKENZIE, *Sinister Street*, 1913

In the glass was a rather plain girl with brown hair and eyes, and a figure well grown but neither particularly graceful nor compact . . . But hope had sprung up, half-suppressed, dubious, irrational, as if a dream had left a sense of prophecy . . . Am I not to be ugly after all?
   Now I'm seventeen I shall begin to fine down . . . But supposing one never did fine down?

ROSAMOND LEHMANN, *Invitation to the Waltz*, 1932

I was sixteen then, and I'm seventeen now, and sometimes I act like I'm about thirteen. It's really ironical because I'm six-feet-two-and-a-half and I have grey hair. I really do.

Holden Caulfield in J. D. SALINGER's *Catcher in the Rye*, 1951

Seventeen—certainly the age when sons react most strongly against their parents.

STEPHEN SPENDER, *World Within World*, 1951

At seventeen, you tend to go in for unhappy love affairs.

FRANÇOISE SAGAN, *Responses*, 1979

When I was seventeen or so,
I scoffed at moneygrubbers.
I had a cold contempt for dough,
And I wouldn't wear my rubbers.

OGDEN NASH, 'Eheu! Fugaces, or What a Difference A Lot of Days Make', *Collected Verse*, 1961

18

It is at this point that self-love changes into self-esteem and that all the passions pertaining to the latter begin to be active.

ROUSSEAU, *Émile*, 1762

In my early years I read very hard. It is a sad reflection but a true one, that I knew almost as much at eighteen as I do now.

SAMUEL JOHNSON to Boswell, 1763, at the age of fifty-four

'Girls should be quiet and modest. The most objectionable part is, that the alteration of manners on being introduced into company is frequently too sudden. They sometimes pass in such very little time from reserve to quite the opposite—to confidence! *That* is the faulty part of the present system. One does not like to see a girl of eighteen or nineteen so immediately up to everything—and perhaps when one has seen her hardly able to speak the year before.'

Mary Crawford in JANE AUSTEN's *Mansfield Park*, 1814

I am very young and perhaps in many, though not in all things, inexperienced, but I am sure, that very few have more real good will and more real desire to do what is fit and right than I have.

QUEEN VICTORIA, *Diary*, 20 June 1837

—How old are you, Marianne?
—And what if I were only eighteen?
—Then you have five or six years to be loved, eight or ten years to love yourself, and the rest to pray to God.

ALFRED DE MUSSET, *Les Caprices de Marianne*, 1851

Said Mr Podsnap to Mrs Podsnap, 'Georgiana is almost eighteen.'
Said Mrs Podsnap to Mr Podsnap, assenting, 'Almost eighteen.'
Said Mr Podsnap then to Mrs Podsnap, 'Really I think we should have some people on Georgiana's birthday.'
Said Mrs Podsnap then to Mr Podsnap, 'Which will enable us to clear off all those people who are due.'

CHARLES DICKENS, *Our Mutual Friend*, 1864–5

This evening, after my bath, I suddenly became so pretty that I spent twenty minutes looking at myself. I am sure that if people could see me tonight, I would be a success. The colour of my complexion is absolutely dazzling and yet delicate and tender; my cheeks have but the faintest tinge of pink; nothing marked but the lines of the lips, the eyebrows and the eyes. Please don't think I am blind when I am looking plain. I see it myself I assure you; and this is the first time I have been looking pretty for a very long while. My painting swallows up everything.

The horrible thing of life is that all must fade, become parchment-like, and perish!

MARIE BASHKIRTSEFF, *Journal*, 1878

Even at eighteen, a mentally voracious young woman cannot live entirely upon scenery.

VERA BRITTAIN, *Testament of Youth*, 1933

For premature adventure one pays an atrocious price. As I told you once, every boy I know who drank at eighteen or nineteen is now safe in his grave.

SCOTT FITZGERALD, letter to his daughter, 1937

Far from the adolescent's vanishing it was only the affluent society —permitting extension of public schooling till eighteen, plus the vast increase in youth going to college—that created the problem of adolescence. Because it meant the postponement of earning a living,

and a postponing of adult sexuality until well past the age when
sexual maturity is reached. Actually the adolescent's estrangement,
the struggle he has in order to find himself and his place in society
—these exist because he has only so recently come to be and not
because he is vanishing.

BRUNO BETTELHEIM, The Children of the Dream, 1969

19

Hark you now! Would any but these boiled brains of nineteen and
two-and-twenty hunt in this weather?

The Shepherd in SHAKESPEARE's The Winter's Tale, 1623

On reaching the gate, I felt pity for my comrade, and waited for him,
and took him on the crupper, saying: 'What would our friends speak
of us tomorrow, if, having left for Rome, we had not pluck enough
to get beyond Siena?' Then the good Tasso said I spoke the truth; and
as he was a pleasant fellow he began to laugh and sing; and in this
way, always singing and laughing, we travelled the whole way to
Rome. I had just nineteen years then and so had the century.

BENVENUTO CELLINI, Autobiography, 1562

Trincavellius Lib. I. consil. 16. had a patient nineteen years of age
extremely melancholy, ob nimium studium, Tarvity & praeceptoris
minas, by reason of overmuch study and his tutor's threats.

ROBERT BURTON, The Anatomy of Melancholy, 1621

Ignorance, intolerance, egotism, self-assertion, opaque perception,
dense and pitiful chuckleheadedness—and an almost pathetic uncon-
sciousness of it all, that is what I was at nineteen and twenty.

MARK TWAIN, letter, 1876

When I left you a new world had just begun to exist for me, the
world of love that was at first drunk with its own desire and
hopeless. Even the journey to Berlin which would otherwise have
charmed me completely, left me cold . . . for the rocks I saw were
not rougher, not harsher than the emotions of my soul, the broad
cities not more full of life than my blood, the tables of the inns not

more overladen and their fare not more indigestible than the stocks
of fantasies that I carried with me.

KARL MARX, letter to his father, November 1837

The passions are the sails of the little ship, you know. And he who in
his twentieth year gives way entirely to his feeling, catches too much
wind and his boat ships too much water and—and he sinks—or
comes to the surface again after all.

VAN GOGH, letter to his brother Theo, November 1881

The time up to the age of eighteen, nineteen is the stage set against
which you create all your work emotionally, socially and in every
sense, until you're finished.

ALAN SILLITOE, 1981

'How old are you, Sally?'
  'Nineteen.'
  'Good God! And I thought you were about twenty-five!'
  'I know. Everybody does.'

Sally Bowles in CHRISTOPHER ISHERWOOD's *Goodbye to
Berlin*, 1939

20

Then come kiss me, sweet and twenty,
Youth's a stuff will not endure.

SHAKESPEARE, *Twelfth Night*, 1623

He that is not handsome at twenty, nor strong at thirty, nor rich at forty, nor wise at fifty, will never be handsome, strong, rich, or wise.

GEORGE HERBERT, *Jacula Prudentum*, 1651

At twenty years of age, the will reigns, at thirty, the wit; and at forty, the judgement.

BENJAMIN FRANKLIN, *Poor Richard's Almanac*, May 1733

Twenty years old, and as yet nothing done for immortality!

SCHILLER, *Don Carlos*, 1787

Being now in her twenty-first year, Maria Bertram was beginning to think matrimony a duty.

JANE AUSTEN, *Mansfield Park*, 1814

When one is twenty, ideas of the outside world and the effect one can have on it take precedence over everything else.

STENDHAL, *Le Rouge et le Noir*, 1830

My *whole life* has been a twenty-year-long battle between poetry and prose, or, if you prefer, between music and law . . . Now I have come to the crossroads and think with terror, 'Whither now?', if I follow my own instinct it will lead me to art, and I believe that is the right path . . . A man can have no more tormenting thought than the prospect of an unhappy, lifeless and superficial future for which he would have only himself to blame.

SCHUMANN, letter to his mother, July 1830

Live as long as you may, the first twenty years are the longest half of your life.

ROBERT SOUTHEY, *The Doctor*, 1834–47

This day I *go out of my* TEENS and become twenty! It sounds so strange to me!

QUEEN VICTORIA, *Diary*, 24 May 1839

I am now in the depths of despondency because of my age. I'm filled with an hysterical despair to think of fifty dull years more. I hate myself and everyone.

RUPERT BROOKE, letter to his mother, 1907, written on the day after his twentieth birthday

Perhaps, until one starts, at the age of seventy, to live on borrowed time, no year will seem again quite so ominous as the one when the formal education ends and the moment arrives to find employment and bear physical responsibility for the whole future. My parents had given me everything they could possibly owe a child and more. Now it was my turn to decide and nobody—not even the Oxford Appointments Board—could help me very far. I was hemmed in by a choice of jails in which to serve my life imprisonment, for how else at twenty can one regard a career which may last as long as life itself, or at the best until that sad moment is reached when the prisoner is released in consideration of good behaviour, with a pension?

GRAHAM GREENE, *A Sort of Life*, 1971

Oh, God, I'm only twenty and I'll have to go on living and living and living.

JEAN RHYS, diary

21

Long-expected one-and-twenty
Ling'ring year, at last is flown;
Pomp and Pleasure, Pride and Plenty,
Great Sir John, are all your own.

Loosened from the minor's tether,
Free to mortgage or to sell,
Wild as wind, and light as feather,
Bid the slaves of thrift farewell.

Call the Bettys, Kates, and Jennys,
Every name that laughs at care;
Lavish of your grandsire's guineas,
Show the spirit of an heir.

SAMUEL JOHNSON, 'A Short Song of Congratulation', sent to
Mrs Thrale for her nephew Sir John Lade, 1780

An O, for ane and twenty Tam!
    An hey, sweet ane and twenty, Tam!
I'll learn my kin a rattlin sang,
    An I saw ane and twenty, Tam.

ROBERT BURNS, 1792

Seventeen, eighteen, nineteen, twenty and then with the waning
months came an ever augmenting sense of the dignity of twenty-
one. Heaven knows I had nothing to 'come into', save the bare
birthday, and yet I esteemed it as a great possession.

CHARLES DICKENS, David Copperfield, 1849–50

A system of hypocrisy, which lasts through whole years, is one
seldom satisfactorily practised by a person of one-and-twenty.

THACKERAY, Vanity Fair, 1847–8

At twenty-one he [Baudelaire] harried his tailor just as, later in life,
he was to drive his publisher to the verge of madness by his insistence
on accuracy of detail, and he demanded fitting after fitting until he
was entirely satisfied. There was always some detail which did not
please him—the tails were too short, or the collar did not fit, or else
the coat did not show a sufficient expanse of shirt front. At last, when
the result was completely to his satisfaction, and he had strutted like a
peacock before the mirror, admiring himself from all angles, he
would turn to the tailor and say with a lordly air: 'Make me a dozen
suits like this!'

Enid Starkie, BAUDELAIRE, 1957

What sort of creed are you coming to? As far as I remember, twenty-one was a devilish age; so intense; and so violently crabwise —this way, that way, and as you say back as much as forward. But it was amazingly exciting too. Never shall I forget arguing with Thoby [Stephen], Lytton, even good old Saxon [Sydney-Turner], hour after hour about good and truth and one's personal emotions as we called them.

VIRGINIA WOOLF, letter to Judith Stephen, 2 December 1939

When a man is tired of life at twenty-one it indicates that he is rather tired of something in himself.

SCOTT FITZGERALD, letter to his daughter, 29 November 1940

It was not until this winter of 1896 when I had almost completed my twenty-second year that the desire for learning came upon me. I began to feel myself wanting in even the vaguest knowledge about many large spheres of thought.

SIR WINSTON CHURCHILL, *My Early Life*, 1930

After the age of twenty, the frame is uncertain, change is hard to pin down, one is less sure of who one is, and other egos with their own court of adherents invade one's privacy with theirs!

V. S. PRITCHETT, *Midnight Oil*, 1971

22

I have heard some people say they should like to be a girl, and a handsome girl too, from thirteen to two-and-twenty, and after that age again to become a man.

JEAN DE LA BRUYÈRE, *Characters*, 1688

> When I was one-and-twenty
>   I heard a wise man say,
> 'Give crowns and pounds and guineas
>   But not your heart away;
> Give pearls away and rubies
>   But keep your fancy free.'
> But I was one-and-twenty,
>   No use to talk to me.

When I was one-and-twenty
    I heard him say again,
'The heart out of the bosom
    Was never given in vain;
'Tis paid with sighs a plenty
    And sold for endless rue.'
And I am two-and-twenty,
    And oh, 'tis true, 'tis true.

A. E. HOUSMAN, *A Shropshire Lad*, 1896

What, still alive at twenty-two
A clean upstanding chap like you?
Sure, if your throat 'tis hard to slit,
Slit your girl's, and swing for it.

HUGH KINGSMILL, parodying A. E. Housman. *The Dawn's Delay*, 1924

I feel as if I have been in the world a thousand years, and I trail my life behind me like an endless scarf. Often I have no desire to live at all. Of course that is foolish. One ought to pull oneself together and shake off such nonsense.

Masha in CHEKHOV's *The Seagull*, 1896

At such time I found the method of Infinite Series; and in summer, 1665, being forced from Cambridge by the plague, I computed the area of the Hyperbola at Boothby, in Lincolnshire, to two and fifty figures by the same method.

SIR ISAAC NEWTON, on his discovery of the binomial theorem at the age of twenty-two, *Commonplace Book 1664–5*

No grey hairs streak my soul,
no grandfatherly fondness there!
I shake the world with the might of my voice,
and walk—handsome,
twentytwoyearold.

MAYAKOVSKY, 'The Cloud in Trousers', 1914

From the date that the doors of his prep-school close
    On the lonely little son
He is taught by precept, insult, and blows
    The Things that are Never Done.

Year after year, without favour or fear,
From seven to twenty-two,
His keepers insist he shall learn the list
Of the things no fellow can do.

RUDYARD KIPLING, 'The Waster', 1930

I confess I am quite disheartened about you. You seem to have no real purpose in life and won't realize at the age of twenty-two that for a man life means work, and hard work if you mean to succeed.

Lady Randolph Churchill to WINSTON CHURCHILL, 1897

So on the wave of post-war enthusiasm I was swept to the House of Commons at just twenty-two years of age as the youngest MP, which I remained for some time.

SIR OSWALD MOSLEY, *My Life*, 1968

At the age of twenty-two I believed myself to be unextinguishable.

SIEGFRIED SASSOON, *Memoirs of a Fox-Hunting Man*, 1927

I compare human life to a large Mansion of Many Apartments, two of which I can only describe, the doors of the rest being as yet shut upon me—The first we step into we call the infant or thoughtless Chamber, in which we remain as long as we do not think—We remain there a long while, and notwithstanding the doors of the second Chamber remain wide open, showing a bright appearance, we care not to hasten to it; but are at length imperceptibly impelled by the awakening of the thinking principle—within us—we no sooner get into the second Chamber, which I shall call the Chamber of Maiden Thought, than we become intoxicated with the light and the atmosphere, we see nothing but pleasant wonders, and think of delaying there forever in delight: However among the effects this breathing is father of is that tremendous one of sharpening one's vision into the heart and nature of Man—of convincing one's nerves that the world is full of Misery and Heartbreak, Pain, Sickness and oppression—whereby this Chamber of Maiden Thought becomes gradually darkened and at the same time on all sides of it many doors are set open—but all dark—all leading to dark passages—We see not the balance of good and evil. We are in a Mist . . .

KEATS, letter to J. H. Reynolds, May 1818

23

I would there were no age between sixteen and three-and-twenty, or that youth would sleep out the rest; for there is nothing in the between but getting wenches with child, wronging the ancientry, stealing, fighting.

The Shepherd in SHAKESPEARE's The Winter's Tale, 1623

'Jane will be quite an old maid soon, I declare. She is almost three and twenty! Lord, how ashamed I should be of not being married before three and twenty!'

Lydia in JANE AUSTEN's Pride and Prejudice, 1813

It may be as well to inform the reader that Miss Fanny Squeers was in her three-and-twentieth year. If there be any one grace or loveliness inseparable from that particular period of life, Miss Squeers may be supposed to have been possessed of it, as there is no reason to suppose that she was a solitary exception to a universal rule. She was not tall like her mother, but short like her father; from the former she inherited a voice of harsh quality; from the latter a remarkable expression of the right eye, something like having none at all.

CHARLES DICKENS, Nicholas Nickleby, 1838–9

Already I feel myself to be a trifle outmoded. I belong to the Beardsley period . . . Cedo junioribus.

MAX BEERBOHM, Diminuendo, 1895

How misplaced is the sympathy lavished on adolescents. There is a yet more difficult age which comes later, when one has less to hope for and less ability to change, when one has cast the die and has to settle into a chosen life without the consolations of habit or the wisdom of maturity, when, as in her own case, one ceases to be une jeune fille un peu folle, and becomes merely a woman, worst of all, a wife. The very young have their troubles, but they have at least a part to play, the part of being very young.

IRIS MURDOCH, The Bell, 1958

At twenty-three, [Einstein] was already the man whom the world later wished, and failed, to understand. He had absolute confidence. He had absolute faith in his own insight. He was set on submerging his personality, for good and all, in the marvels of the natural world.

EINSTEIN, in C. P. Snow, *Variety of Men*, 1967

24

> How soon hath Time the subtle thief of youth,
>   Stol'n on his wing my three and twentieth year!
>   My hasting days fly on with full career,
>   But my late spring no bud or blossom sheweth.

MILTON, Sonnet VII, written on his twenty-fourth birthday, 9 December 1632

About the first age [Adolescence] there is no doubt, but all wise men agree that it lasts till the twenty-fifth year; and because up to this time our soul is concerned with the growth and improvement of the body (whence many and great transformations take place in our person), the rational part of us has not yet come to perfect discretion.

DANTE, *The Banquet*, 1304–8

I am neither young enough nor old enough to be in love.

HORACE WALPOLE, letter to the Hon. H. S. Conway, 1741

> Now fare thee well, old twenty-three,
> No powers, no arts can thee retain;
> Eternity will roll away,
> And thou wilt never come again.
>
> And welcome thou, young twenty-four,
> Thou bringest to men of joy and grief;
> Whatever thou bringest in sufferings sour,
> The heart in faith will hope relief.

CARLYLE, 1808

> A man o' four-an'-twenty
> that 'asn't learned of a trade—

Beside 'Reserve' agin' him—'e'd
better be never made.
RUDYARD KIPLING, 'Back to the Army Again', *Barrack Room
Ballads*, 1896

It's something to become a bore,
And more than that, at twenty-four.
SIR JOHN BETJEMAN, 'The Wykehamist', *Mount Zion*, 1932

Twenty-four years remind the tears of my eyes.
(Bury the dead for fear they walk to the grave in labour.)
In the groin of the natural doorway I crouched like a tailor
Sewing a shroud for a journey
By the light of the meat-eating sun.
Dressed to die, the sensual strut begun.
With my red veins full of money,
In the final direction of the elementary town
I advance for as long as forever is.
DYLAN THOMAS, October 1938

Nobody should try to be a writer until he reaches twenty-four years
of age.
JAMES THURBER, letter to Robert Reilly, 23 February 1961

Pitt was Prime Minister at four-and-twenty, and that precedent has
ruined half our young politicians.
TROLLOPE, *Phineas Finn*, 1869

Since I was twenty-four . . . there never was any vagueness in my
plans or ideas as to what God's work was for me.
FLORENCE NIGHTINGALE writing in 1857 about her realiza-
tion in 1844 that her vocation lay in hospitals among the sick

Let me tell you, that when I was twenty-four I'd have thought
myself an old has-been if I hadn't already clocked up a husband, a
divorce, and a seven-year-old son!
LIANE DE POUGY, the French courtesan, *My Blue Notebooks*,
1979

25

Youth is the only season for enjoyment, and the first twenty-five years of one's life are worth all the rest of the longest life of man, even though those five-and-twenty be spent in penury and contempt, and the rest in the possession of wealth, honours, respectability.

GEORGE BORROW, *The Romany Rye*, 1857

Ah, what should I be at fifty
   Should nature keep me alive,
If I find the world so bitter
   When I am but twenty-five?

TENNYSON, 'Maud', 1855

No person shall be a representative who shall not have attained to the age of twenty-five years.

Constitution of the United States, 1788

I never object to a certain degree of disputatiousness in a young man from the age of seventeen to that of four or five and twenty, provided I find him always arguing on one side of the question.

COLERIDGE, *Biographia Literaria*, 1817

A poet who has not produced a good poem before he is twenty-five, we may conclude cannot, and never will do so.

WORDSWORTH, 1820

Everyone has talent at twenty-five, the difficulty is to have it at fifty.

DEGAS

My dear, infant prodigies stay seventeen until they are twenty-five.

Max Jacob, talking about the novelist RAYMOND RADIGUET

In October 1879 Rimbaud reached his twenty-fifth birthday, and his friends began to notice a change in him. Delahaye describes him at the end of 1879 as having considerably calmed down and subdued. He seemed to have lost all taste for alcohol and for excitement, and there was, in the expression of his eyes, something gentle and spiritual once more . . . Rimbaud told Delahaye that his days of wandering were over—it was as if he had suddenly come to new decisions—and he told him of his ambitions for the future. He talked now of it as if he saw some direction and pattern in it, but he did not mention any of his old interests—neither history, philosophy, nor literature. 'And what about literature', Delahaye asked suddenly. 'Oh! I never think of that now!' he answered gruffly. Then he hastily changed the subject . . .

                          Enid Starkie, RIMBAUD, 1961

She had reached a great age—for it quite seemed to her that at twenty-five it was late to reconsider; and her most general sense was a shade of regret that she had not known earlier.

          HENRY JÁMES, *The Wings of the Dove*, 1902

The effective, moving, vitalizing work of the world is done between the ages of twenty-five and forty.

          SIR WILLIAM OSLER, address to Johns Hopkins University 1905

Brain weight peaks at about twenty-five, and the number of critical cells, after a period of constancy from birth to the early twenties, declines sharply to the nineties. Each day of our adult lives more than 100,000 nerve cells die and nerve cells are never, of course, replaced.

          LORD ROTHSCHILD, 'Too Old?', Melchett Lecture, 1972

I was twenty-five, and too old to be unusual.

          JAMES D. WATSON, *The Double Helix*, 1968, at the age when he won the Nobel Prize for his part in discovering the structure of the DNA molecule

26

His Highness [Philip of Spain] was middle-aged, being twenty-six, and it seemed difficult to find a prince of the age she required, for if she chose a husband of fifty he would be too old to hope for posterity, and men declined and grew old at fifty or sixty, which age

very few passed . . . His Highness had already been married, had a
son of eight and was a prince of so stable and settled a character that
he was no longer young, for nowadays a man nearly thirty was
considered as old as men formerly were at forty.

> Simon Renard, the Spanish Imperial Ambassador, in a report to
> the Emperor Charles V, describing his talks with Queen Mary I
> about her marriage, October 1553

'Invite him to dinner, Emma, and help him to the best of the fish and
the chicken, but leave him to chuse his own wife. Depend upon it, a
man of six or seven-and-twenty can take care of himself.'

> Mr Knightley in JANE AUSTEN'S Emma, 1816

Colonel Osborne was always so dressed that no-one ever observed
the nature of his garments, being no doubt well aware that no man
after twenty-five can afford to call special attention to his hat, his
cravat or his trousers.

> TROLLOPE, He Knew He was Right, 1869

> It happened to Lord Lundy then
> As happens to so many men:
> Towards the age of twenty-six,
> They shoved him into politics.

> HILAIRE BELLOC, Cautionary Tales, 1907

But haven't I, at twenty-six, reached the age when one should begin
to learn?

> RUPERT BROOKE, letter to Edward Marsh, April 1914

It was not long after that that everybody was twenty-six. During the
next two or three years all the young men were twenty-six years old.
It was the right age apparently for that time and place . . . If they
were young men they were twenty-six. Later on, much later on,
they were twenty-one and twenty-two.

> GERTRUDE STEIN, The Autobiography of Alice B. Toklas, 1933

When a woman reaches twenty-six in America, she's on the slide. It's
downhill all the way from then on. It doesn't give you a tremendous
feeling of confidence and well-being.

> LAUREN BACALL, interview in the Observer magazine, January,
> 1979

27

He was but seven-and-twenty, an age at which many men are not quite common—at which they are hopeful of achievement, resolute in avoidance, thinking that Mammon shall never put a bit in their mouths and get astride their backs, but rather that Mammon, if they have anything to do with him, shall draw their chariot.

GEORGE ELIOT, *Middlemarch*, 1871–2

The happy time is, of course, gone by in which we saw every object haloed by youth; now comes the unpleasant recognition of miserable reality, which (thank God) I try to spare myself as far as possible by the exercise of my imagination.

SCHUBERT, letter to his brother Ferdinand, 1824

Mme Ranevsky: You're twenty-six or twenty-seven years old, but you're still like a schoolboy in a prep school!. . . You ought to be a man, at your age you ought to understand people who are in love. And you ought to be able to love   to fall in love!
Trofimov: What is she saying?
Mme Ranevsky: 'I'm above love!' You're not above love, you're daft, as our Feers would say. Not to have a mistress at your age!

CHEKHOV, *The Cherry Orchard*, 1904

Her body was going meaningless, going dull and opaque, so much insignificant substance. It made her feel immensely depressed and hopeless. What hope was there? She was old, old at twenty-seven, with no gleam and sparkle in the flesh. Old through neglect and denial, yes, denial.

D. H. LAWRENCE, *Lady Chatterley's Lover*, 1928

It was about then [1920] that I wrote a line which certain people will not let me forget: 'She was a faded but still lovely woman of twenty-seven.'

SCOTT FITZGERALD, *Early Success*, 1937

With what elevated, what triumphant feelings unseen and unnoticed by the world my life is filled! I swear I will do something that the ordinary person does not do. I feel leonine strength in my soul, and I perceive clearly my transition from childhood spent in school exercises, to a young age.

GOGOL, letter to V. Zhukovsky, June 1836

The time was one of intellectual intoxication. My sensations resembled those one has after climbing a mountain in a mist, when, on reaching the summit, the mist suddenly clears, and the country becomes visible for forty miles in every direction. For years I had been endeavouring to analyse the fundamental notions of mathematics, such as order and cardinal numbers. Suddenly, in the space of a few weeks, I discovered what appeared to be definitive answers to the problems which had baffled me for years. And in the course of discovering these answers, I was introducing a new mathematical technique, by which regions formerly abandoned to the vaguenesses of philosophers were conquered for the precision of exact formulae. Intellectually, the month of September 1900 was the highest point of my life. I went about saying to myself that now at last I had done something worth doing, and I had the feeling that I must be careful not to be run over in the street before I had written it down.

BERTRAND RUSSELL, *Autobiography*, i, 1967

28

> And oh! I shall find how, day by day,
>   All thoughts and things look older;
> How the laugh of pleasure grows less gay,
>   And the heat of friendship colder.

W. M. PRAED, 'Twenty-eight and Twenty-nine', *Poems*, 1823–30

The year hastens to its close. What is it to me? What I am, that is all that affects me. That I am twenty-eight or eight or fifty-eight years old is as nothing. Should I mourn that the spring flowers are gone, that the summer fruit has ripened, that the harvest is reaped, that the snow has fallen?

EMERSON, *Journals*, 28 December 1831

Jack decides, as a worthy fellow of twenty always will decide, that mere external rank and convenience are nothing; the dignity of mind is all in all. I argue as every reasonable man of twenty-eight, that this is poetry in part, which a few years will mix pretty largely with prose.

CARLYLE, letter to his father, 2 April 1824

From 1921–9 I was a thousand Socrates. I was very conscious of being 'the young Sartre' but I remembered Goethe's dictum: 'whoever is not famous at twenty-eight must give up any dreams of glory'.

JEAN-PAUL SARTRE, *Diaries*, written at the age of thirty-four

Adieu! I must now write to the king of France, compose a solo for flute, make up a poem for Voltaire, alter some army regulations, and do a thousand things!

FREDERICK THE GREAT OF PRUSSIA, letter to a friend, just after his accession in 1740

I have written two grand concertos, and also a quintet for hautboy, clarinet, corno, bassoon, and pianoforte, [K. 452] which was received with extraordinary applause. I consider it myself to be the best thing I ever wrote in my life. How I wish you could have heard it, and how beautifully it was executed! But, to tell you the truth, towards the close I was quite worn out with incessant playing, and I think it is much to my credit that my audience were not so also.

MOZART, letter to his father, April 1784

I feel like a well-appointed laboratory of the soul—myself, my home, my life—in which none of the vitally fecund or destructive, explosive experiments has yet begun. I like the shape of the bottles, the colours of the chemicals. I collect bottles, and the more they look like alchemist bottles the more I like them for their eloquent forms.

ANAÏS NIN, *Journal*, 1931

Until I was twenty-eight I had a kind of buried self who didn't know she could do anything but make white sauce and diaper babies. I didn't know I had any creative depths. All I wanted was a little piece of life, to be married, to have children . . . The surface cracked when I was about twenty-eight. I had a psychotic break and tried to kill myself.

ANNE SEXTON, when asked why it was not until she was almost thirty that she began to write. *Paris Review* no. 52, 1968

29

What art thou, trow?—
  A man worth praise, perfay.—
This is thy thirtieth year of wayfaring.—
  'Tis a mule's age—Art thou a boy still?—Nay.

FRANÇOIS VILLON, 'The Dispute of the Heart and Body of
François Villon', c. 1451, trans. Swinburne

Methinks I have outlived myself, and begin to be weary of the sun; I
have shaken hands with delight, in my warm blood and canicular
days, I perceive I do anticipate the vices of age; the world to me is but
a dream or mock-show, and we all therein but Pantalones and
Anticks, to my severer contemplations.

SIR THOMAS BROWNE, *Religio Medici*, 1635

The knell of my thirtieth year has sounded; in three or four years my
youth will be as a faint haze on the sea, an illusive recollection.

GEORGE MOORE, *Confessions of a Young Man*, xii, 1888

If ye live enough befure thirty ye won't care to live at all afther fifty.

Mr Dooley (FINLEY PETER DUNNE), *Casual Observations*, 1900

Any person under the age of thirty, who, having any knowledge of
the existing social order, is not a revolutionist, is an inferior.

BERNARD SHAW, *The Revolutionist's Handbook*, 1903

Michelangelo's fame was grown so great that in the year 1503, when
he was twenty-nine years of age, Julius II sent for him to come and
build his tomb.

GIORGIO VASARI, *The Lives of the Painters*, 1550, 1568

I am sick of society. I need solitude, isolation. My feelings are dried
up, and I am bored with public display. I am tired of glory at
twenty-nine; it has lost its charm; and there is nothing left for me but
complete egotism.

NAPOLEON, letter to Joseph Bonaparte, 25 July 1798

Thirteen winters' revolving frosts had seen her opening every ball of credit which a scanty neighbourhood afforded, and thirteen springs shewn their blossoms, as she travelled up to London with her father, for a few weeks' annual enjoyment of the great world. She had the remembrance of all this, she had the consciousness of being nine-and-twenty to give her some regrets and some apprehensions; she was fully satisfied of being still quite as handsome as ever, but she felt her approach to the years of danger, and would have rejoiced to be certain of being properly solicited by baronet-blood within the next twelvemonth or two.

JANE AUSTEN, *Persuasion*, 1818

The day before the sea closed over my own Shelley, he said to Marianne: 'If I die tomorrow, I have lived to be older than my father. I am ninety years of age.'

Mary Shelley, letter to Mrs Gisborne, 10 September 1822, of SHELLEY in his thirtieth year

You remember I used to say I wanted to die at thirty—well, I'm now twenty-nine and the prospect is still welcome. My work is the only thing that makes me happy—except to be a little tight—and for those two indulgences I pay a big price in mental and physical hangovers.

SCOTT FITZGERALD, letter to Maxwell Perkins, 1925

I have never admitted that I am more than twenty-nine, or thirty at the most. Twenty-nine when there are pink shades, thirty when there are not.

Mrs Erlynne in OSCAR WILDE's *Lady Windermere's Fan*, 1892

When you come to write my epitaph, Charles, let it be in these delicious words, 'She had a long twenty-nine.'

Rosalind in JAMES BARRIE's *Rosalind*, 1914

30

Not childhood alone, but the young man till thirty, never feels practically that he is mortal.

CHARLES LAMB, *Essays of Elia*, 1823

At thirty man suspects himself a fool;
Knows it at forty, and reforms his plan;
At fifty chides his infamous delay,
Pushes his prudent purpose to resolve;
In all his magnanimity of thought
Resolves; and re-resolves; then dies the same.

EDWARD YOUNG, *Love of Fame*, 1725–8

At thirty a man should know himself like the palm of his hand, know the exact number of his defects and qualities, know how far he can go, foretell his failures—be what he is. And above all accept these things.

ALBERT CAMUS, 30 July 1945, from *Carnets*, 1942–51

Beware, O my dear young men, of going rotten.
It's so easy to follow suit;
people in their thirties, and the older ones, have gotten bad inside, like fruit
that nobody eats and nobody wants, so it rots but is not forgotten.

D. H. LAWRENCE, 'Beware, O My Dear Young Men', *Pensées*, 1929

Strength of body, and that character of countenance which the French term a *physionomie*, women do not acquire before thirty, any more than men.

MARY WOLLSTONECRAFT, *Vindication of the Rights of Woman*, 1792

No man,
Till thirty, should perceive there's a plain woman.

BYRON, *Don Juan*, XIII, iii, 1819–24

For a young man a woman of thirty has irresistible attractions.

BALZAC, *The Woman of Thirty*, 1832

At twenty a man is rash in love, and again, perhaps, at fifty; a man of middle-age enamoured of a young girl is capable of sublime follies. But the man of thirty who loves for the first time is usually the embodiment of cautious discretion. He does not fall in love with a violent descent, but rather lets himself gently down, continually testing the rope.

ARNOLD BENNETT, *Anna of the Five Towns*, 1902

The whole person of Arkady's uncle, with its aristocratic elegance, had preserved a youthful shapeliness and that soaring quality, up and away from the earth, which usually disappears when a man has turned thirty.

TURGENEV, *Fathers and Sons*, 1862

The age of thirty is, for the working man, just the beginning of a period of some stability, and as such, one feels young and full of energy. But, at the same time, a period of life is passed, which makes one melancholy, thinking some things will never come back. And it is no silly sentimentalism to feel a certain regret. Well, many things really begin at the age of thirty, and certainly all is not over then. But one doesn't expect from life what one has already learned it cannot give, but rather one begins to see more and more clearly that life is only a kind of sowing time, and the harvest is not here.

VAN GOGH, letter to his brother Theo, 8 February 1883

Peter, in our boyish years,
Thirty to forty,
When Cupid was still the Christ of Love's religion,
time stood on its hands.

ROBERT LOWELL, 'Our Afterlife I', *Day by Day*, 1978

Although Bertha Young was thirty she still had moments like this when she wanted to run instead of walk, to take dancing steps on and off the pavement, to bowl a hoop, to throw something in the air and catch it again, or to stand still and laugh at—nothing—at nothing, simply.

What can you do if you are thirty and, turning the corner of your own street, you are overcome, suddenly, by a feeling of bliss —absolute bliss!—as though you'd suddenly swallowed a bright piece of that afternoon sun and it burned in your bosom, sending out a little shower of sparks into every particle, into every finger and toe? . . .

Oh, is there no way you can express it without being 'drunk and disorderly'? How idiotic civilisation is! Why be given a body if you have to keep it shut up in a case like a rare, rare fiddle?

KATHERINE MANSFIELD, *Bliss*, 1920

> Soon, soon the flesh
> The grave cave ate will be
> At home on me
>
> And I a smiling woman.
> I am only thirty.
> And like the cat I have nine times to die.

SYLVIA PLATH, 'Lady Lazarus', 1965, written after her third suicide attempt

31

Any man over thirty identifies his youth with the worst fault he thinks he is capable of.

CESARE PAVESE, *Diary*, September 1938

After thirty, a man wakes up sad every morning excepting perhaps five or six, until the day of his death.

EMERSON, *Journals*, 1834

I never sought to please till past thirty years old, considering that matter hopeless.

SAMUEL JOHNSON to Henry Thrale

Now I know that in thus turning Conservative with years, I am going through the normal cycle of change and travelling in the orbit of men's opinions. I submit to this as I would submit to gout or grey hair as a concomitant of growing age or else of failing animal heat; but I dare say it is deplorably for the worst.

ROBERT LOUIS STEVENSON, *Virginibus Puerisque*, 1881

Since death (to be precise) is the true end and purpose of our life, I have made it my business over the past few years to get to know this true, this best friend of man so well that the thought of him not only holds no terrors for me but even brings me great comfort and peace of mind.

MOZART, letter to his father, 4 April 1787

I shall be thirty-one . . . My youth is gone like a dream; and very little use have I ever made of it. What have I done these last thirty years? Precious little . . .

CHARLOTTE BRONTË, letter to Ellen Nussey, 24 March 1847

This lady was now somewhat past thirty, an era at which, in the opinion of the malicious, the title of old maid may with no impropriety be assumed.

HENRY FIELDING, *Tom Jones*, 1749

32

At our age, when you have reached, not merely by the process of thought but with your whole being and your whole life, an awareness of the uselessness and impossibility of seeking enjoyment; when you feel that what seemed like torture has become the only substance of life—work and toil—then searchings, anguish, dissatisfaction with yourself, remorse etc.—the attributes of youth—are inappropriate and useless.

TOLSTOY, letter to B. N. Chicherin, January 1860

If you would not be forgotten, as soon as you are dead and rotten, either write things worth reading, or do things worth the writing.

BENJAMIN FRANKLIN, *Poor Richard's Almanac*, 1738

The atrocious crime of being a young man, which the honourable gentleman has, with such spirit and decency, charged upon me, I shall neither attempt to palliate nor deny; but content myself with wishing that I may be one of those whose follies cease with their youth, and not of those who continue ignorant in spite of age and experience.

> WILLIAM PITT the Elder replying to Walpole, 6 March 1741

I am thinking what a veneration we used to have for Sir William Temple because he might have been Secretary of State at fifty; and here is a young fellow hardly thirty in that appointment.

> SWIFT, *Journal to Stella*, 11 November 1710, referring to the appointment of Henry St John (later Lord Bolingbroke) as Secretary of State at the age of thirty-two

I think there cannot be a mortal on this earth less commended than I; but I will not change. I will not reform. I have already erased, corrected, blotted out or suppressed so many things in myself that I am weary of it. Everything has its end, and I think I am now a big enough boy to consider my education complete. Now I have other things to think about. I was born with all the vices. I have radically suppressed some, and kept the rest on a starvation diet. God alone knows the martyrdoms I have undergone in this psychological training school; but now I give up. That path leads to the grave, and I want to live through three or four more books.

> FLAUBERT, letter to Louise Colet, January 1834, written while he was working on *Madame Bovary*

for the 4 years more I have of life up to 35 no Im what am I at all Ill be 33 in September will I what O well look at that Mrs Galbraith shes much older than me I saw her when I was out last week her beautys on the wane

> Molly Bloom in JAMES JOYCE's *Ulysses*, 1922

33

It is not, I confess, an unlawful prayer to desire to surpass the days of our Saviour, or wish to outlive that age wherein He thought fittest to die; yet if (as Divinity affirms) there shall be no grey hairs in Heaven, but all shall rise in the perfect state of men, we do but outlive those perfections in this world, to be recalled unto them by a greater miracle in the next, and run on here but to be retrograde hereafter.

SIR THOMAS BROWNE, *Religio Medici*, 1642

I am thirty-three,—the age of the good sans-culotte Jesus; an age fatal to revolutionists.

CAMILLE DESMOULINS, when asked his age by the Tribunal two days before being guillotined, 1794. Carlyle notes that his real age was thirty-four

He [Julius Caesar] saw a statue of Alexander the Great in the Temple of Hercules at Cadiz, and was heard to sigh—peeved, it appears, that at an age when Alexander had already conquered the world, he himself had not done anything at all remarkable.

SUETONIUS, *Lives of the Caesars*, c. 121

The truth is, I do indulge myself a little the more in pleasure, knowing that this is the proper age of my life to do it; and, out of my observation that most men that do thrive in the world do forget to take pleasure during the time that they are getting their estate, but reserve that till they have got one, and then it is too late for them to enjoy it.

SAMUEL PEPYS, *Diary*, 10 March 1666

It is three minutes past twelve.—' 'Tis the middle of the night by the castle clock', and I am now thirty-three!

> *Eheu, fugaces, Posthume, Posthume,*
> *Labuntur anni,—*

but I don't regret them so much for what I have done, as for what I *might* have done.

> Through life's road, so dim and dirty,
> I have dragged to three-and-thirty.
> What have these years left to me?
> Nothing—except thirty-three.

BYRON, *Diary*, 22 January 1821

> O that I now could, by some chymic art,
> To sperm convert my vitals and my heart,

That at one thrust I might my soul translate,
And in the womb myself regenerate:
There steeped in lust, nine months I would remain:
Then boldly —— my passage out again.

EARL OF ROCHESTER, 1680

When I began to write it (The Tropic of Cancer) in New York I was
thirty-three, and you know why I left it till then? I would have been
lingering still, but my wife told me one day, 'If you haven't begun
your life's work at the age of thirty-three, which is the age Christ
died, you will never do it', and that lodged in my mind. Then one
day she said, 'Why do you continue doing this office work—give it
all up', and I said 'What would we live on?', and she said 'I'll take care
of you, don't worry—I'll devote myself to take care of you', and she
did, you know.

HENRY MILLER, February 1978

Good Christ, a Jewish man with parents alive is a fifteen-year-old
boy, and will remain a fifteen-year-old boy till *they die*! . . . Doctor!
Doctor! Did I say fifteen? Excuse me, I meant ten! I meant five! I
meant zero! A Jewish man with his parents alive is half the time a
helpless *infant*! Spring me from this role I play of the smothered son
in the Jewish joke! Because it's beginning to pall a little, at thirty-
three!

PHILIP ROTH, *Portnoy's Complaint*, 1969

34

And inasmuch as the master of our life, Aristotle, was aware of this
arch of which we are speaking, he seemed to maintain that our life
was no other than a mounting and a descending, wherefore he says in
that wherein he treats of Youth and Age, that youth is no other than
the growing of life. It is hard to say where the highest point of this
arch is, because of the inequality spoken of above; but in the majority
I take it to be somewhere between the thirtieth and the fortieth year.
And I believe that in those of perfect nature it would be in the
thirty-fifth year.

DANTE, *The Banquet*, 1304–8

THE OXFORD BOOK OF AGES

Stella this day is thirty-four,
(We won't dispute a year or more)
However Stella, be not troubled,
Although thy size and years are doubled,
Since first I saw thee at sixteen,
The brightest virgin on the green;
So little is thy form declined;
Made up so largely in thy mind . . .

> SWIFT, 'Stella's Birthday', 13 March 1718/19

Four years and thirty, told this very week,
Have I been now a sojourner on earth,
And yet the morning gladness is not gone
Which then was in my mind.

> WORDSWORTH, *The Prelude*, Book VI, 1805

Trelawny: Byron said to me the other morning: 'I was reminded by a letter from my sister that I was thirty-four; but I felt at that time that I was twice that age. I must have lived fast.'
Shelley: The mind of man, his brain, and nerves; are a truer index of his age than the calendar, and that may make him seventy.
   When Shelley at a later date said he was ninety, he was no doubt thinking of the wear and tear of his own mind.

> E. J. Trelawny, *Records of Shelley, Byron and the Author*, 1878

. . . But warm, eager, living life—to be rooted in life—to learn, to desire to know, to feel, to think, to act. That is what I want. And nothing less. That is what I must try for.

> KATHERINE MANSFIELD, *Journal* 14 October 1922. She died three months later

At the age of thirty-four, she had at length become aware why fate had chosen her for a remarkable position . . . A complete transformation both of her outer and of her inner life occurred in the Tuileries. The woman who for twenty years had never been able to hear an envoy patiently to the end of his say, who had never read a letter or a book attentively, who had cared for nothing beyond cards, sport, the fashion, and other trifles, now changed her writing-table into that of a chancellor, her room into a diplomat's cabinet.

> Stefan Zweig, MARIE ANTOINETTE, 1933

One evening, in January 1883, Gauguin came home from the Rue Carcel and told his wife that he had handed in his resignation to Galichon and Calzado. Mette stared at him aghast. 'I've handed in my resignation,' he repeated calmly. 'From now on I shall paint every day.'

Henri Perruchot, GAUGUIN, 1963

35

Midway upon the journey of our life
I found myself within a forest dark.
DANTE, *The Divine Comedy*, c. 1307, trans. Longfellow

Oft in danger, yet alive,
We are come to thirty-five:
Long may better years arrive,
Better years than thirty-five.
Could philosophers contrive
Life to stop at thirty-five,
Time his hours should never drive
O'er the bounds of thirty-five.
High to soar and deep to dive,
Nature gives at thirty-five.
Ladies, stock and tend your hive,
Trifle not at thirty-five:
For howe'er we boost and strive,
Life declines from thirty-five:
He that ever hopes to thrive
Must begin by thirty-five:
And all who wisely wish to wive
Must look on Thrale at thirty-five.
SAMUEL JOHNSON, for Mrs Thrale, 1776

Alas! at thirty-five to be still preparing for something!
TURGENEV, *Rudin*, 1856

Sensible people get the greater part of their own dying done during their own lifetime. A man at five and thirty should no more regret

not having had a happier childhood than he should regret not having
been born a prince of the blood.

SAMUEL BUTLER, *The Way of All Flesh*, 1903

Mozart died in his six-and-thirtieth year. Raphael at the same age.
Byron only a little older. But all these had perfectly fulfilled their
missions; and it was time for them to depart, that other people might
still have something to do in a world made to last a long while.

GOETHE to Johann Eckermann, 11 March 1828

The shock, for an intelligent writer, of discovering for the first time
that there are people younger than himself who think him stupid is
severe. Especially if he is at an age (thirty-five to forty-two) when his
self-confidence is easily shattered. The seventh lustre is such a
period, a menopause for artists, a serious change of life. It is the
transition from being a young writer, from being potentially Byron,
Shelley, Keats, to becoming a stayer, a Wordsworth, a Coleridge, a
Landor. It would seem that genius is of two kinds, one of which
blazes up in youth and dies down, while the other matures, like
Milton or Goethe's, through long choosing, putting out new branches every seven years.

CYRIL CONNOLLY, *Enemies of Promise*, 1938

> This, no song of an ingenue,
>   This, no ballad of innocence;
> This, the rhyme of a lady who
>   Followed ever her natural bents.
> This, a solo of sapience,
> This, a chantey of sophistry,
>   This, the sum of experiments—
> I loved them until they loved me.

DOROTHY PARKER, 'Ballade at Thirty-Five'

That fatal ten years between thirty-five and forty-five is neither
youth nor age, the beauty of youth declined and the beauty of age not
yet arrived.

KATHERINE ANNE PORTER, letter to Glenway Westcott, 1959

I'd like to go on being thirty-five for a long time.

MARGARET THATCHER, May 1979

36

And thus ends all that I doubt I shall ever be able to do with my own eyes in the keeping of my Journal, I being not able to do it any longer, having done now so long as to undo my eyes almost every time that I take a pen in my hand; and, therefore, resolve from this time forward, to have it kept by my people in long-hand, and must therefore be contented to set down no more than is fit for them and all the world to know; or, if there be anything, which cannot be much, now my amours to Deb. are past, and my eyes hindering me in almost all other pleasures, I must endeavour to keep a margin in my books open, to add, here and there, a note in short-hand in my own hand.

And so I betake myself to that course, which is almost as much as to see myself go into my grave: for which, and all the discomforts that will accompany my being blind, the good God prepare me!

> SAMUEL PEPYS, the last entry in his diary, 8 May 1669. He could still read just before his death, at the age of seventy

Let me tell you, a woman labours under many disadvantages who tries to pass for a girl at six and thirty.

> Mrs Candour in SHERIDAN's *The School for Scandal*, 1777

> My days are in the yellow leaf;
>   The flowers and fruits of love are gone;
> The worm, the canker, and the grief
>   Are mine alone!

> BYRON, 'On This Day I Complete my Thirty-Sixth Year', January 1824; he died in April

All has now been said. If at my age one has still not known happiness, pure and unalloyed, Nature will not henceforth allow one to put the cup to one's lips; white hairs may not approach it.

> BALZAC to Madame Hanska, 1836

When I was thirteen years old, my Uncle Samuel Ripley one day asked me, 'How is it, Ralph, that all the boys dislike you and quarrel with you, whilst the grown people are fond of you?'

Now I am thirty-six and the fact is reversed—the old people suspect and dislike me, and the young love me.

EMERSON, *Journals*, 30 September 1839

Nothing is consistent, nothing is fixed or certain, in my life. By turns I resemble and differ; there is no living creature so foreign to me that I cannot be sure of approaching. I do not yet know, at the age of thirty-six, whether I am miserly or prodigal, temperate or greedy . . . or rather, being suddenly carried from one to the other extreme, in this very balancing I feel that my fate is being carried out.

ANDRÉ GIDE, *Journal*, 24 August 1905

Never trust a woman who wears mauve, whatever her age may be, or a woman over thirty-five who is fond of pink ribbons. It always means that they have a history.

OSCAR WILDE, *The Picture of Dorian Gray*, 1891

I was asked the other day. 'What are you doing nowadays?' 'I'm busy growing older,' I answered, 'it's a whole-time job.'

PAUL LÉAUTAUD, *Journal*, 31 December 1907

37

While I am writing this essay, I find myself about middle age, computing life according to the calculation of the Royal Psalmist. From the point therefore where I now am, I can most impartially judge of youth and old age . . . I must fairly acknowledge that in my opinion the disagreement between young men and old is owing rather to the fault of the latter than of the former. Young men, though keen and impetuous, are usually very well disposed to receive the counsels of the old, if they are treated with gentleness . . . But old men forget in a wonderful degree their own feelings in the early part of life . . .

JAMES BOSWELL, 'The Hypochondriack', *London Magazine*, June 1778

Thirty-seven years, more than half the life of man are run out.
—What an atom, an animalcule I am!—The remainder of my days I
shall rather decline, in sense, spirit, and activity. My season for
acquiring knowledge is past. And yet I have my own and my
children's fortunes to make. My boyish habits and airs are not yet
worn off.

JOHN ADAMS, *Diary*, 1840

I felt a strange need to cry out something about myself, which would
impose upon them the truth of my existence—My age for instance!
That is impressive, the age of a man! That summarizes all his life.
This maturity of his has taken a long time to achieve. It has grown
through so many obstacles conquered, so many serious illnesses
cured, so many griefs appeased, so many despairs overcome, so
many dangers unconsciously passed . . . The age of a man, that
represents a good load of experience and memories! In spite of
decoys, jerks and ruts, you have continued to plod like a horse
drawing a cart. And now, because of a determined convergence of
lucky chances, there you are. You are thirty-seven.

SAINT-EXUPÉRY, *Letter to a Hostage*, 1943

Every day I attach less and less importance to the intellect. Every day
I realize more that it is only by other means that a writer can regain
something of our impressions, reach, that is, a particle of himself, the
only material of art. What the intellect restores to us under the name
of the past is not the past . . .

MARCEL PROUST, soon after receiving the inspiration for
writing *A la Recherche du Temps Perdu*, January 1909

For, being seized with a fever, he made his will, and having
confessed, he ended his course on the same day that he was born, that
is, Good Friday, being thirty-seven years of age. They placed at the
head of the room in which he lay, the picture of the Transfiguration,
which he had finished for the Cardinal de' Medici, and the sight of
the dead body and the living work filled all with grief.

Giorgio Vasari on the death of RAPHAEL in 1520. *The Lives of
the Painters*, 1550, 1568

Ah! that fatal thirty-seven, which reminds me of Byron, greater even
as a man than as a writer. Was it experience that guided the pencil of

Raphael when he painted the palaces of Rome? He, too, died at thirty-seven.

DISRAELI, *Coningsby*, 1844

Against a setting of dust-coloured brocade, and opposite my brother, sat [Wyndham] Lewis . . . At the end of dinner, in the quietness, he first made a calculation with a pencil on an old match box and then, with the usual yellowing cigarette stub clamped to his upper lip while he spoke . . . said firmly, but in a carefully lowered voice: 'Remember! I'm thirty-seven till I pass the word around!'

SIR OSBERT SITWELL, *Laughter in the Next Room*, 1948

38

Well, here I am at thirty-eight,
Well, I certainly thought I'd have longer to wait.
You just stop in for a couple of beers,
And gosh, there go thirty-seven years.
Well, it has certainly been fun,
But I certainly thought I'd have got a lot more done . . .

After thirty-seven years I find myself the kind of man that anybody
    can sell anything to,
And nobody will tell anything to.

OGDEN NASH, 'Not George Washington's, not Abraham Lincoln's, but Mine', *Collected Verse*, 1961

And so, dear mother, this scribble must end, as others have done. Tomorrow, I believe, is my eight-and-thirtieth birthday! You were then young in life; I had not yet entered it. Since then—how much! how much! They are in the land of silence, whom we once knew. . . . A few years more and we too shall be with them in eternity. Meanwhile it is this *Time* that is ours: let us be busy with it and work, work, 'for the Night cometh'.

CARLYLE, letter to his mother, 3 December 1833

The day after my birthday; in fact, I'm thirty-eight, well, I've no doubt I'm a great deal happier than I was at twenty-eight.

VIRGINIA WOOLF, *A Writer's Diary*, 26 January 1920

By the bye, as I must leave off being young, I find many Douceurs in being a sort of Chaperon for I am put on the Sofa near the fire & can drink as much wine as I like.

JANE AUSTEN, letter to her sister Cassandra, 6 November 1813

The further I advance in life, the more convinced I am of the necessity of that principle of wisdom which befits our nature: enjoy what lies in your own hands.

DELACROIX, letter to Frederic Villot, 26 September 1840

I think that to so many what happens is the death of ambition in the conventional sense. That great driving motor that prods you and exasperates you and brings out the worst qualities in you for about twenty years is beginning to be a bit moth-eaten and tired. I find that I'm altogether much quieter, I think; I don't love music any less; but there's not the excess of energy that I used to spend in enthusiasm and in intoxication. I feel much freer than I've ever been in my life.

COLIN DAVIS, conductor, interview in *The Observer*, April 1966

I'd decided that I would make my life my argument. I would advocate the things I believed in terms of the life I lived and what I did.

ALBERT SCHWEITZER, who sailed to Africa in 1913 to set up his hospital at Lambaréné

39

When forty winters shall besiege thy brow,
And dig deep trenches in thy beauty's field,
Thy youth's proud livery, so gazed on now,
Will be a tattered weed, of small worth held . . .

SHAKESPEARE, Sonnet 2

The first forty years of life furnish the text, while the remaining thirty supply the commentary.

> SCHOPENHAUER, 'Counsels and Maxims', *Parerga and Paralipomena*, 1851

I began to realize that for two years of my life I had been drawing on resources I did not possess, that I had been mortgaging myself physically and spiritually up to the hilt.

> SCOTT FITZGERALD, *The Crack-Up*, 1936

Benvenuto Cellini said that a man should be at least forty years old before he undertakes so fine an enterprise as that of setting down the story of his life. He said also that an autobiographer should have accomplished something of excellence. Nowadays nobody who has a typewriter pays attention to the old master's quaint rules. I myself have accomplished nothing of excellence except a remarkable and, to some of my friends, unaccountable expertness at hitting empty ginger-beer bottles with small rocks at a distance of thirty paces. Moreover, I am not yet forty years old. But the grim date moves towards me apace; my legs are beginning to go, things blur before my eyes, and the faces of the rose-lipped maids I knew in my twenties are misty as dreams.

> JAMES THURBER, 'Preface to a Life', 25 September 1933

My thirty-ninth birthday. A good year. I have begotten a fine daughter, published a successful book, drunk 300 bottles of wine, and smoked 300 or more Havana cigars. I have got back to soldiering among friends . . . I get steadily worse as a soldier with the passage of time, but more patient and humble—as far as soldiering is concerned. I have about £900 in hand and no grave debts except to the Government; health excellent except when impaired by wine; a wife I love, agreeable work in surroundings of great beauty. Well that is as much as one can hope for.

> EVELYN WAUGH, *Diaries*, 28 October 1942

He'd met her shortly before his fortieth birthday. He was at that time of life when a man becomes aware of his plumbing, when he wakes in the night to sounds of protest from within—joints creaking, intestines in a riot.

> PAUL BAILEY, *Old Soldiers*, 1980

The fact that Bradman was about to undertake a tour of England in his fortieth year meant far less in 1948 than it might mean today. The age of forty was decidedly not considered the veteran stage in English cricket thirty years ago, even though for most Australians it may have been 'old'.

Irving Rosenwater, SIR DONALD BRADMAN, 1978

40

Forty is a terrible age.

CHARLES PÉGUY, *Notre Jeunesse*, 1910

Forty is the old age of youth; fifty is the youth of old age.

VICTOR HUGO (1802–85)

We don't understand life any better at forty than at twenty, but we know it and admit it.

JULES RENARD (1864–1910)

Be wise with speed;
A fool at forty is a fool indeed.

EDWARD YOUNG, *Love of Fame*, 1725–8

At twenty years of age, the will reigns; at thirty, the wit; and at forty, the judgement.

BENJAMIN FRANKLIN, *Poor Richard's Almanac*, 1758

The man who is not a socialist at twenty has no heart, but if he is still a socialist at forty he has no head.

ARISTIDE BRIAND (1862–1932)

To hold the same views at forty as we held at twenty is to have been stupefied for a score of years, and take rank, not as a prophet, but as an unteachable brat, well birched and none the wiser. It is as if a ship's captain should sail to India from the Port of London; and having brought a chart of the Thames on deck at his first setting out, should obstinately use no other for the whole voyage.

ROBERT LOUIS STEVENSON, *Virginibus Puerisque*, 1881

Better one bite, at forty, of Truth's bitter rind,
Than the hot wine that gushed from the vintage of twenty!
JAMES RUSSELL LOWELL, 'Two Scenes from The Life of
Blondel', *Collected Poems*, 1903

No one becomes forty without incredulity and a sense of outrage.
CLIFFORD BAX, October 1925

Nobody feels well after his fortieth birthday
But the convalescence is touched by glory . . .
PETER PORTER, 'Returning', *English Subtitles*, 1981

The English kill all their poets off by the time they're forty.
D. H. LAWRENCE

We have no permanent habits until we are forty. Then they begin to
harden, presently they petrify, then business begins. Since forty I
have been regular about going to bed and getting up—and that is one
of the main things.
MARK TWAIN, 'Seventieth Birthday', 1950

It may be said that married men of forty are usually ready and
generous enough to fling passing glances at any specimen of moderate beauty they may discern by the way.
THOMAS HARDY, *Far From the Madding Crowd*, 1874

When an artist allows himself to have two or three successes while
young, still under thirty, he can be sure that the public will then
grow tired of him. They begin to resent having showered applause
on such a whippersnapper . . . if the artist has the strength to stand
up to this turn of the tide and go ahead on his own path, he'll be safe
by the time he's forty.
VERDI, letter to Countess Maffei, 1876

I am very well, and only wish I were not so lazy; but I hope and
believe one is less so from forty to fifty, if one lives, than at any other
time of life. The loss of youth ought to operate as a spur to one to live
more by the head, when one can live less by the body.
MATTHEW ARNOLD, letter to his sister, 25 June 1859

He was, I trowe, a twenty wynter oold,
And I was fourty, if I shal seye sooth;
But yet I hadde alwey a coltes tooth.
Gat-tothed I was, and that bicam me weel;
I hadde the prente of seinte Venus seel.
As help me God! I was a lusty oon,
And faire, and riche, and yong, and wel bigon . . .

> CHAUCER, *The Wife of Bath's Prologue*, speaking of her fifth husband, *c.* 1387

I am forty now, and forty years is a lifetime; it is extremely old age. To go on living after forty is unseemly, disgusting, immoral! Who goes on living after forty? give me a sincere and honest answer! I'll tell you—fools and rogues.

> DOSTOYEVSKY, *Notes from the Underground*, 1864

She told him that the next day would be her own fortieth birthday. He replied, 'If you were hit by a car today the papers would say that a young woman was run over. If it happens tomorrow, they'll make it: "A fortyish woman . . ."'

> Herbert Lottman, ALBERT CAMUS: A Biography, 1979

I have a bone to pick with Fate.
Come here and tell me, girlie,
Do you think my mind is maturing late,
Or simply rotted early?

> OGDEN NASH, 'Lines on Facing Forty', *Collected Verse*, 1961

41

Every man over forty is a scoundrel.

> BERNARD SHAW, *The Revolutionist's Handbook*, 1903

Our youth began with tears and sighs,
With seeking what we could not find
Our verses were all threnodies,
In elegiacs still we whined,
Our ears were deaf, our eyes were blind,

We sought and knew not what we sought.
We marvel, now we look behind:
Life's more amusing than we thought!

ANDREW LANG, 'Ballade of Middle Age', 1885

As far as the Olympic Games is concerned, there are few competitors
over the age of forty, except in shooting, riding, fencing, marathon
and long-distance-walking contests. In most cases, the individual is
discouraged by his declining performance and withdraws from
competition, but even if he persists the age handicap leads to early
elimination in national heats.

R. J. SHEPHERD, *The Fit Athlete*, 1978

'But Mr Weston is almost an old man. Mr Weston must be between
forty and fifty.'

'Which makes his good manners the more valuable. The older a
person grows, Harriet, the more important it is that their manners
should not be bad—the more glaring and disgusting any loudness,
or coarseness, or awkwardness becomes. What is passable in youth,
is detestable in later age.'

JANE AUSTEN, *Emma*, 1816

Let me assure you, dear, the Viennese let neither grey hair nor
stooped shoulders get in the way of making new conquests. Every
day I see handsome young men helping their much older lovers into
their coaches. In this worldly capital, women under thirty-five are
regarded as immature. A woman can't even hope to make a lasting
impression on people until she's over forty.

LADY MARY WORTLEY MONTAGU, letter to Lady Rich, 20
September 1716

The *Sunday Express* today published a most extraordinary paragraph
to the effect that I am really forty-one instead of thirty-nine, and
hinted that I had faked my age in the reference books. The awful
thing is that it is true. Now I feel apprehensive and shy, as one does
when one is in disgrace. Honor is being very sweet and loyal about it
. . . I told her she would be a widow two years earlier.

CHIPS CHANNON, *Diaries*, 19 June 1938

My heart throbs; I tremble, I cannot stop sighing. Is it possible at my age? What is happening is the same as when I was twenty-five. Yet I feel confidence and (incredibly) serenely hopeful.

> CESARE PAVESE, *Diary*, March 1950, a few months before he committed suicide

It was only in my forties that I started feeling young.

> HENRY MILLER, 'On Turning Eighty', *Sextet*, 1977

42

> This day, whate'er the Fates decree,
> Shall still be kept with joy by me;
> This day then, let us not be told,
> That you are sick, and I grown old,
> Nor think on our approaching ills,
> And talk of spectacles and pills;
> Tomorrow will be time enough
> To hear such mortifying stuff.
>
> SWIFT, 'Stella's Birthday', 13 March 1726/7

> A boy may still detest age,
>   But as for me, I know,
> A man has reached his best age
>   At forty-two or so.
>
> R. C. LEHMANN, 'Middle Age', *Anni Fugaces*, 1901

Where are the fine verses of Solon, to the effect that the orator has not the harmony of thought and speech until from about forty-two to fifty-six years of age?

> EMERSON, *Journals*, 1863

. . . Urgent necessity to recover possession of oneself. But can one still make resolutions when one is over forty? I live according to twenty-year-old habits. Did I know what I was doing at twenty? When I made the resolution to look at everything, never to prefer

myself to anything and always to give preference to what differed
most from me . . .

ANDRÉ GIDE, Journal, 14 January 1912

When I disengaged myself as above mentioned from private business, I flattered myself that by the sufficient though moderate
fortune I had acquired, I had secured leisure during the rest of my
life, for philosophical studies and amusements . . .

BENJAMIN FRANKLIN in 1748, Autobiography 1771–1790

Mr Salteena was an elderly man of 42 and was fond of asking peaple
to stay with him. He had quite a young girl staying with him of 17
named Ethel Monticue. Mr Salteena had dark short hair and mus-
tache and wiskers which were very black and twisty. He was middle
sized and he had very pale blue eyes. He had a pale brown suit but on
Sundays he had a black one and he had a topper every day as he
thorght it more becoming.

DAISY ASHFORD, The Young Visiters, 1919, written at the age
of nine

Instead of bewailing a lost youth, a man nowadays begins to
wonder, when he reaches my ripe age of forty-two, if ever his past
will subside and be comfortably by-gone. Doing over these poems
makes me realize that my teens and my twenties are just as much me,
here and now and present, as ever they were, and the pastness is only
an abstraction. The actuality, the body of feeling, is essentially alive
and here.

D. H. LAWRENCE, Foreword to Juvenilia, 1928

43

I broke the back of life yesterday and started downhill toward old
age. This fact has not produced any effect on me that I can detect.

MARK TWAIN, letter to his mother on his forty-third birthday,
1878

Am I not in a curiously *unnatural* state of mind in this way—that at
forty-three, instead of being able to settle to my middle-aged life like

a middle-aged creature, I have more instincts of youth about me than when I was young, and am miserable because I cannot climb, run, or wrestle, sing, or flirt—as I was when a youngster because I couldn't sit writing metaphysics all day long. Wrong at both ends of life . . .

RUSKIN, letter to Dr John Brown, 11 January 1862

If age, not experience, is the standard, he [Kennedy] went on, then a maturity test excluding 'from positions of trust and command all those below the age of forty-four would have kept Jefferson from writing the Declaration of Independence, Washington from commanding the Continental Army, Madison from fathering the Constitution . . . and Christopher Columbus from even discovering America'. (He wisely struck out the one other name I had on the list, that of Jesus of Nazareth.)

Theodore Sorensen, KENNEDY, 1965, describing J. F. Kennedy's presidential election campaign in 1960

The one thing I regret is that I will never have time to read all the books I want to read.

FRANÇOISE SAGAN, Responses, 1979

Mrs Bubb was gay and free, fair, fat and forty-three,
And blooming as a peony in buxom May,
The toast she long had been of Farringdon Within,
And she filled the better half of a one-horse chay.

Anon., Modern Street Ballads, 1888

She may very well pass for forty-three
In the dusk, with a light behind her!

W. S. GILBERT, Trial by Jury, 1875

Her age shall be no secret to the reader, though to her most intimate friends, even to Mr Browne, it had never been divulged. She was forty-three, but carried her years so well, and had received such gifts from nature, that it was impossible to deny that she was a beautiful woman . . . She did not fall in love, she did not wilfully flirt, she did not commit herself; but she smiled and whispered, and made confidences, and looked out of her own eyes into men's eyes as though there might be some mysterious bond between her and them—if only mysterious circumstances would permit it.

TROLLOPE, The Way We Live Now, 1874–5

His face, when he was alone, changed so that it was hardly recogniz-able. The muscles of his mouth and cheeks, otherwise obedient to his will, relaxed and became flabby. Like a mask the look of vigour, alertness, and amiability, which now for a long time had been preserved only by constant effort, fell from his face, and betrayed an anguished weariness instead. The tired, worried eyes gazed at objects without seeing them; they became red and watery. He made no effort to deceive even himself; and of all the dull, confused, rambling thoughts that filled his mind he clung to only one: the single, despairing thought that Thomas Buddenbrooks, at forty-three years, was an old, worn-out man.

THOMAS MANN, *Buddenbrooks*, 1901

44

What is it to grow old?
Is it to lose the glory of the form,
The lustre of the eye?
Is it for beauty to forego her wreath?
—Yes, but not this alone.
          . . .
It is—last stage of all—
When we are frozen up within, and quite
The phantom of ourselves,
To hear the world applaud the hollow ghost
Which blamed the living man.

MATTHEW ARNOLD, 'Growing Old', 1867

It has been thought by some that life is like the exploring of a passage that grows narrower and darker the farther we advance, without a possibility of ever turning back, and where we are stifled for want of breath at last. For myself, I do not complain of the greater thickness of the atmosphere as I approach the narrow house. I felt it more formerly, when the idea alone seemed to suppress a thousand rising hopes, and weighed upon the pulses of the blood. At present I feel rather a thinness and want of support, I stretch out my hand to some object and find none, I am too much in a world of abstraction; the naked map of life is spread out before me, and in the emptiness and desolation I see Death coming to meet me. In my youth I could not behold him for the crowd of objects and feelings, and Hope always stood between us, saying, 'Never mind that old fellow!'

WILLIAM HAZLITT, 'On the Fear of Death', *Table Talk*, 1822

Say the woman is forty-four.
Say she is five-seven-and-a-half.
Say her hair is stick colour.
Say her eyes are chameleon.
Would you put her in a sack and bury her,
Suck her down into the dumb dirt?
    Some would,
    If not, time will.

ANNE SEXTON, 'Hurry Up Please It's Time', *The Death Note-books*, 1975

I do not wish to die and even should like to reach deep old age, but I should refuse to become young once again and live my whole life from the beginning. Once is enough.

TCHAIKOVSKY, letter to Anna Merkling, 1884

I would hate to die, ever, because every year I have a better time fishing and shooting. I like them as much as when I was sixteen and now I've written good enough books so that I don't have to worry about that I would be happy to fish and shoot and let somebody else lug the ball for a while.

ERNEST HEMINGWAY, letter to Hadley Mowrer, 23 July 1942

Mailer never had a particular age—he carried different ages within him like different models of his experience: parts of him were eighty-one years old, fifty-seven, forty-eight, thirty-six, nineteen, et cetera, et cetera—he now went back abruptly from fifty-seven to thirty-six.

NORMAN MAILER at forty-four, *The Armies of the Night*, 1968

45

Forty-five is the age of recklessness for many men, as if in defiance of the decay and death waiting with open arms in the sinister valley at the bottom of the inevitable hill.

JOSEPH CONRAD, *Victory*, 1915

Well, who wants to be young anyhow, any idiot born in the last
   forty years can be young, and besides forty-five isn't really old, it's
   right on the border;

At least, unless the elevator's out of order.

OGDEN NASH, 'Let's Not Climb the Washington Monument
Tonight', *Collected Verse*, 1961

There is no fool like an old fool,
   Yet fools of middling age
Can seldom teach themselves to reach
   True folly's final stage.

Their course of love mounts not above
   Some five-and-forty years,
Though God gave men threescore and ten
   To scald with foolish tears.

ROBERT GRAVES, 'Fools', *Poems 1970–2*

At forty-five,
what next, what next?
At every corner,
I meet my Father,
my age, still alive.

ROBERT LOWELL, 'Middle Age', *For the Union Dead*, 1965

You will think it my fate either to find or to imagine some lady of forty-five, very unprejudiced and philosophical, who has entered deeply into the best and selectest spirit of the age, with enchanting manners, and a disposition rather to like me, in every town that I inhabit. But certainly such this lady is.

SHELLEY, letter to Leigh Hunt, 5 April 1820

I stand unspeakably alone, today—which happens to be my birthday—more than ever.

IBSEN, 20 March 1873

Top-level achievement in sports declines by age forty-five to a level
not reached until age seventy in the sciences. There seems some basis
for accepting the general view that physical capacities develop and
decline earliest, whilst psychological capacities develop later.

JAMES E. BIRREN, *The Psychology of Ageing*, 1964

It is absurd to believe that one can conclude a life of battles at the age
of forty-five . . . I was leader of the revolution and chief of the
Government at thirty-nine. Not only have I not finished my job, but
I feel that I have not begun it.

MUSSOLINI, *Autobiography*, 1928

Comment ça va, middle age?
Qu'est ce que tu as, middle age?
Autumn winds begin to blow
And so
I'd better unbend my mind to you
Though—you know
I'm not resigned to you,
More relaxation,
More ease,
More time for snoozing,
What consolation
Are these
For those amusing
Pleasures I'm losing?
Shall I survive this decade
Or shall I merely fade out,
Done for—played out?
What are your designs for the final page?

NOEL COWARD, 'Middle Age', *The Girl Who Came to Dinner*,
1963

46

Old age begins at forty-six years, according to the common opinion.

CICERO, *De Senectute, c.* 44 BC

I shall never forget with what regret he spoke of the rude reply he made to Dr Barnard, on his saying that men never improved after the age of forty-five. 'That is not true, Sir,' said Johnson. 'You, who perhaps are forty-eight, may still improve, if you will try: I wish you would set about it; and I am afraid', he added, 'there is great room for it'; and this was said in rather a large party of ladies and gentlemen at dinner . . .

Frances Reynolds, *Recollections of* DR JOHNSON, 1829. See also p.97

Control of life is what one should learn now: its economic management. I feel cautious, like a poor person, now I am forty-six.

VIRGINIA WOOLF, *Diary*, 18 March 1928

She's six-and-forty, and I wish nothing worse to happen to any woman.

SIR ARTHUR WING PINERO, *The Second Mrs Tanqueray*, 1893

I have drunk of the wine of life at last, I have known the thing best worth knowing, I have been warmed through and through, never to grow quite cold again till the end.

EDITH WHARTON, in love for the first time, *Journal*, May 1908

His name was George F. Babbit. He was forty-six years old now, in April, 1920, and he made nothing in particular, neither butter nor shoes nor poetry, but he was nimble in the calling of selling houses for more than people could afford to pay.

His large head was pink, his brown hair thin and dry. His face was babyish in slumber, despite his wrinkles and the red spectacle-dents on the slopes of his nose. He was not fat but he was exceedingly well fed; his cheeks were pads, and the unroughened hand which lay helpless upon the khaki-colored blanket was slightly puffy. He seemed prosperous, extremely married and unromantic; and altogether unromantic appeared this sleeping-porch, which looked on one sizable elm, two respectable grass-plots, a cement driveway, and a corrugated iron garage. Yet Babbit was again dreaming of the fairy child, a dream more romantic than scarlet pagodas by the silver sea.

SINCLAIR LEWIS, *Babbit*, 1922

. . . I suppose that I shall have to die beyond my means.

OSCAR WILDE, who died at the age of forty-six in 1900

47

When I reflect, with serious sense,
    While years and years run on,
How soon I may be summoned hence—
    There's cook a-calling John.

Our lives are built so frail and poor
    On sand and not on rocks,
We're hourly standing at Death's door—
    There's someone double-knocks.

All human days have settled terms,
    Our fates we cannot force;
This flesh of mine will feed the worms—
    They've come to lunch of course.

And when my body's turned to clay,
    And dear friends hear my knell,
Oh, let them give a sigh and say—
    I hear the upstairs bell.

THOMAS HOOD, 'A Few Lines on Completing Forty-Seven',
Domestic Didactics By An Old Servant, 1839

Then, for the first time in my life, I reflected on my past and cursed all my deeds and the fifty years that I would shortly attain: O horror! with the weight of forty-seven years upon me . . . In my old age I find no pleasure save in the memories which I have of the past.

CASANOVA, *Memoirs*, 1821

Swift was then about forty-seven, at an age when vanity is strongly excited by the amorous attentions of a young woman.

SWIFT in Samuel Johnson, *Lives of the Poets*, 1779–81

When one is twenty yes, but at forty-seven Venus may rise from the sea, and I for one should hardly put on my spectacles to have a look.

THACKERAY, letter to Dr John Brown, November 1858

How short life is! How much I have still to do, to think and to say! We keep putting things off and meanwhile death lurks round the corner.

TCHAIKOVSKY, *Journal*, September 1887

I'm forty-seven now. Up to a year ago I tried deliberately to pull the wool over my eyes . . . so that I shouldn't see the realities of life . . . and I thought I was doing the right thing. But now—if you only knew! I lie awake, night after night, in sheer vexation and anger —that I let time slip by so stupidly during the years when I could have had all the things from which my age now cuts me off.

Vania in CHEKHOV's *Uncle Vania*, 1900

And then I am forty-seven: yes; and my infirmities will of course increase. To begin with my eyes. Last year, I think, I could read without spectacles; would pick up a paper and read it in a tube; gradually I found I needed spectacles in bed; and now I can't read a line (unless held at a very odd angle) without them.

VIRGINIA WOOLF, *A Writer's Diary*, 10 September 1929

> Take a maiden tender—her affection raw and green,
>   At very highest rating,
>   Has been accumulating
> Summers seventeen—summers seventeen.
>   Don't, beloved master,
>   Crush me with disaster,
> What is such a dower to the dower I have here?
>   *My* love unabating
>   Has been accumulating
> Forty-seven year—forty-seven year!
>
> Ruth in W. S. GILBERT's *The Pirates of Penzance*, 1800

Ah me: the years that have gone since in the pride of young manhood
I first went forth to the fight. I grow old and tired but must always
lead on.

> SIR ERNEST SHACKLETON on his third and last expedition, to
> the South Atlantic. Diary, January 1922

48

What changes time makes in the mind. A letter from an old friend
raises in me emotions very different from those which I felt when we
used to talk to one another formerly by every post or carrier. As we
come forward into life we naturally turn back now and then upon the
past. I now think more upon my schooldays than I did when I had
just broken loose from a master. Happy is he that can look back upon
the past with pleasure.

> SAMUEL JOHNSON, letter to Edmund Hector, 11 November
> 1756

I lately thought no man alive
Could ere improve past forty-five,
    And ventured to assert it.
The observation was not new,
And seemed to me so just and true
    That none could contravert it.

'No Sir', says Johnson, ' 'tis not so,
'Tis your mistake, and I can show
    An instance, if you doubt it.
You, who perhaps are forty-eight,
May still improve, 'tis not too late:
    I wish you'd set about it' . . .

> DR BARNARD, later Bishop of Limerick, replying to Dr John-
> son's assertion that he could improve after the age of forty-
> eight. Frances Reynolds, *Recollections of Dr Johnson*, 1829. See
> also p. 94

Pardon that for a barren passion's sake
Although I have come close on forty-nine,

I have no child, I have nothing but a book,
Nothing but that to prove your blood and mine.

> W. B. YEATS, 'Introductory Rhymes' from *Responsibilities*,
> 1914

Isn't it a great pity that the nearer I come to that age when, so they tell
us, the only happiness to be enjoyed lies in the quietness that results
from being unable to move, I can find some well-being only in
emotion; for that's synonymous with agitation and disturbance,
which are attributes of youth.

> DELACROIX, letter to George Sand, 12 September 1846

I am forty-eight, and my hair is turning white.

> BALZAC to Madame Hanska, 1847

Excursion to the beechwood. Some hundreds of young people have
collected there; they sing melodies belonging to the time when I was
young, thirty years ago. They play the games and dance the dances
of my youth. Melancholy overcomes me, and suddenly my whole
past life unrolls before the eyes of my spirit. I can survey the path I
have traversed, and feel dazzled. Yes, it will soon end. I am old, and
the path descends to the grave. I cannot restrain my tears—I am old.

> AUGUST STRINDBERG, *Extracts from My Diary*, 1897. Three
> years later he married, and thereafter wrote many of his greatest
> plays

I have so much to write about myself, that I don't know where to
begin. I have been going through such a continuous process of
experience (in the last eighteen months) that I can hardly find words
for them. How can I attempt to describe such a tremendous crisis! I
see everything in such a new light—I am so constantly on the move;
sometimes it would hardly surprise me to discover that I'd suddenly
taken on a new body (like Faust in the last scene). My thirst for life is
keener than ever, and I find the 'habit of living' even sweeter than
before.

> MAHLER to Bruno Walter, 1909, trans. Hans Gal

So little done, so much to do.

> Dying words (according to Dr Jamieson) of CECIL RHODES,
> 26 March 1902

49

The body is at its best between the ages of thirty and thirty-five: the
mind at its best about the age of forty-nine.

ARISTOTLE, *Rhetoric, c.* 322 BC

When I was as you are now, towering in the confidence of twenty-
one, little did I suspect that I should be at forty-nine, what I now am.

SAMUEL JOHNSON, letter to Bennett Langton, 1758

I go to a concert, party, ball—
What profit is in these?
I sit alone against the wall
And strive to look at ease.
The incense that is mine by right
They burn before Her shrine;
And that's because I'm seventeen
And She is forty-nine.

I cannot check my girlish blush,
My colour comes and goes.
I redden to my finger-tips,
And sometimes to my nose.
But She is white where white should be,
And red where red should shine.
The blush that flies at seventeen
Is fixed at forty-nine.

RUDYARD KIPLING, 'My Rival', *Departmental Ditties*, 1885

O, wad some pawky power
Gie me a gowden giftie,
I'd like to stop at forty-nine,
But pontificate like fifty.

OGDEN NASH, 'The Calendar-Watchers *or* What's So Wonderful About Being a Patriarch?', *Collected Verse*, 1961

Our hidalgo was getting on for fifty, of strong constitution, lean,
thin-faced, an early riser and a lover of hunting.

CERVANTES, *Don Quixote*, 1604

My musical memory . . . until my fiftieth year, was prodigious; but since then, I have been conscious of a growing weakness. I begin to feel an uncertainty; something like a nervous dread often takes possession of me on stage . . . I often fear lest memory betray me into forgetfulness of a passage, and that I may unconsciously change it . . . This sense of uncertainty has often inflicted upon me tortures only to be compared with those of the Inquisition, while the public listening to me imagines that I am perfectly calm.

ARTUR RUBINSTEIN, *Autobiography*, 1982

By 1965, the year in which I became Leader of the Opposition, I was almost fifty, and again I was being told that I could not go on indefinitely at the pace I had been living for the past twenty years if I wanted to carry heavy political burdens in the future . . . We began to talk boats and how to sail them. And that is how it all began . . .

EDWARD HEATH, *Sailing*, 1975

50

At fifty, everyone has the face he deserves.

GEORGE ORWELL, *Notebook*, April 1949

When have I last looked on
The round green eyes and the long wavering bodies
Of the dark leopards of the moon?
All the wild witches those most noble ladies,
For all their broomsticks and their tears,
Their angry tears, are gone.
The holy centaurs of the hills are vanished;
I have nothing but the embittered sun;
Banished heroic mother moon and vanished,
And now that I have come to fifty years
I must endure the timid sun.

W. B. YEATS, 'Lines Written in Dejection', 1919

Mens Sana? O at last; from twenty years
of the annual mania, thirty of adolescence:
this crown. And am I still *in corpore sano*?
Some mornings now my studies wane by eleven,
afternoons by three. The print, its brain,
clouds in mid-chapter, just as I will go—
two score and ten . . . less than common expectation?
All the new years already last? Old times,
when death was nothing, death the dirty crown
on a sound fingernail—ephemeral,
though avant-garde? Now I can tan on my belly
without impatience, almost hear out old people,
live off the family chronicle—the swallow
scents out the kinship, dares swoop me from my nest.

ROBERT LOWELL, 'Sound Mind, Sound Body', *Notebook*, 1970

Every age has its admirers, ladies. While you, perhaps, are trading among the warmer climes of youth, there might be some to carry on a useful commerce in the frozen latitudes beyond fifty.

Honeywood in GOLDSMITH's *The Good-Natur'd Man*, 1768

Love is lame at fifty years.

THOMAS HARDY, 'The Revisitation', *Time's Laughingstocks*, 1909

Fifty today, old lad?
Well, that's not doing so bad:
All those years without
Being really buggered about.
The next fifty won't be so good,
True, but for now—touch wood—
You can eat and booze and the rest of it,
Still get a lot of the best of it,
While the shags with fifty or so
Actual years to go
Will find most of them tougher . . .

KINGSLEY AMIS, 'Ode to Me', *Collected Poems 1944–1979*

The years between fifty and seventy are the hardest. You are always being asked to do things, and yet you are not decrepit enough to turn them down.

T. S. ELIOT, 23 October 1950

A man past fifty should never write a novel.

THACKERAY, in Trollope, *Life of Thackeray*, 1879

I don't suppose anyone will be much interested in what I have to say this time and it may be the last novel I'll ever write [*The Last Tycoon*] but it must be done now because, after fifty, one is different. One can't remember emotionally, I think, except about childhood, but I have a few more things left to say.

SCOTT FITZGERALD, letter to his wife Zelda, 11 October 1940

Really, one should ignore one's fiftieth birthday. As anyone over fifty will tell you, it's no age at all. All the same, it is rather sobering to realize that one has lived longer than Arnold of Rugby, or Porson, the eighteenth-century professor of Greek. It's hard not to look back and wonder why one hasn't done more, or forward and wonder what, if anything, one will do in the future.

PHILIP LARKIN, *The Listener*, 17 August 1972

If there is a single date to which we can attach the establishment of Stalin's power as absolute dictator, it is that of his fiftieth birthday on December 21, 1929.

Ronald Hingley, JOSEPH STALIN: *Man and Legend*, 1974

51

Past fifty, we learn with surprise and a sense of suicidal absolution
that what we intended and failed
could never have happened—
and must be done better.

ROBERT LOWELL, *Day by Day*, 1977

Every age has its springs which give it movement; but man is always the same. At ten years old he is led by sweetmeats; at twenty by a mistress; at thirty by pleasure; at forty by ambition; at fifty by avarice; after that what is left for him to run after but wisdom?

ROUSSEAU, *Moral Letters*, 1757

In the fifty-second year of my age, after the completion of an arduous and successful work, I now propose to employ some moments of my leisure in reviewing the simple transactions of a private and literary life. Truth, naked, unblushing truth, the first virtue of more serious history, must be the sole recommendation of this personal narrative . . .

EDWARD GIBBON, *Autobiography*, 1788

I am ageing fast, I am tired of life, I thirst for quietness and a rest from all these vanities, emotions, disappointments, etc. etc. It is natural for an old man to think of a prospective dirty hole called a grave.

TCHAIKOVSKY, 1891. He died two years later

She was not old yet. She had just broken into her fifty-second year. Months and months of it were still untouched. June, July, August! Each still remained almost whole, and, as if to catch the falling drop, Clarissa (crossing to the dressing-table) plunged into the very heart of the moment, transfixed it, there—the moment of this June morning on which was the pressure of all the other mornings, seeing the glass, the dressing-table, and all the bottles afresh, collecting the whole of her at one point (as she looked into the glass), seeing the delicate pink face of the woman who was that very night to give a party; of Clarissa Dalloway; of herself.

VIRGINIA WOOLF, *Mrs Dalloway*, 1925

Everything got better after I was fifty. I wrote my best books. I walked Pillar, starting from Buttermere, which I'm told no fell walker of advanced years should attempt.

A. J. P. TAYLOR, interview in the *Evening Standard*, March 1982

One night at dinner, Palgrave fussily told the waiter to be particularly careful of the port because 'the old gentleman' was fastidious about his wine. When the waiter was gone, Tennyson asked, 'Do you mean me by the old gentleman?' At fifty-one he was sensitive about his age . . .

R. B. Martin, TENNYSON: *The Unquiet Heart*, 1980

It is my experience that towards middle age a man has three choices: to stop writing altogether, to repeat himself with perhaps an increasing skill of virtuosity, or by taking thought to adapt himself to middle age and find a different way of writing.

T. S. ELIOT, memorial lecture on W. B. Yeats, June 1940

Old people are more interesting than young. One of the particular points of interest is to observe how after fifty they revert to the habits, mannerisms and opinions of their parents, however wild they were in youth.

EVELYN WAUGH, letter to Nancy Mitford, 29 October 1963

52

We wondered, Shakespeare, that thou went'st so soon
From the world's stage to the grave's tiring room.

                    'JM' on the death of SHAKESPEARE, April 1616

When all the fiercer passions cease
    (The glory and disgrace of youth);
When the deluded soul, in peace,
    Can listen to the voice of truth;
When we are taught in whom to trust,
    And how to spare, to spend, to give,
(Our prudence kind, our pity just,)
    'Tis then we rightly learn to live.

                    GEORGE CRABBE, 'Reflection', 1807

Every day, and all day long, I ask myself this question—or rather
this question asks itself of me: Shall I find it hard to die?
    I do not think that death is particularly hard for those who most
loved life. On the contrary.

                    ANDRÉ GIDE, *Journal*, 1922

Just why Lincoln took to whiskers at this time nobody seemed to
know. A girl in New York State had begged him to raise a beard. But
something more than her random wish guided him. Herndon,
Whitney, Lamon, Nicolay, Hay, heard no explanation from him as
to why after fifty-two years with a smooth face he should now
change.
    Would whiskers imply responsibility, gravity, a more sober and
serene outlook on the phantasmagoria of life? Perhaps he would
seem more like a serious farmer with crops to look after, or perhaps a
church sexton in charge of grave affairs. Or he might have the look
of a sea-captain handling a ship in a storm on a starless sea. Anyhow,
with whiskers or without, he would be about the same-sized target.

                    Carl Sandburg, ABRAHAM LINCOLN, 1939

Mrs Candour: But surely now, her sister *is*, or *was*, very handsome.
Crabtree: Who? Mrs Evergreen? O Lord! She's six-and-fifty if she's
    an hour!

Mrs Candour: Now positively you wrong her; fifty-two or fifty-three is the utmost—and I don't think she looks more.
Sir Benjamin: Ah! there's no judging by her looks, unless one could see her face.

SHERIDAN, The School for Scandal, 1777

I cannot see a likely young creature without impatiently considering her chances for, say, fifty-two. O, you mysterious girls, when you are fifty-two we shall find you out; you must come into the open then.

J. M. BARRIE, The Little White Bird, 1902

I refuse to admit that I'm more than fifty-two even if that does make my sons illegitimate.

LADY ASTOR

53

When I was young my teachers were the old.
I gave up fire for form till I was cold.
I suffered like a metal being cast.
I went to school to age to learn the past.

Now I am old my teachers are the young.
What can't be moulded must be cracked and sprung.
I strain at lessons fit to start a suture.
I go to school to youth to learn the future.

ROBERT FROST, 'What Fifty Said'

Last winter I went down to my native town, where I found the streets much narrower and shorter than I thought I had left them, inhabited by a new race of people, to whom I was very little known. My playfellows were grown old, and forced me to suspect that I was no longer young.

SAMUEL JOHNSON, letter to Joseph Baretti, 20 July 1762

Happiness.
I have been happy for two years.

It mayn't be over yet, but I want to write it down
before it gets spoiled by pain—which is the chief
thing pain can do in the inside life: spoil the lovely
things that had got in there first.
Happiness can come in one's natural growth and not
queerly, as religious people think. From fifty-one to fifty-three
I have been happy, and would like to remind others
that their turn can come too. It is the only
message worth giving.

E. M. FORSTER, *Commonplace Book*, 1932

He could hardly be called old at the age of fifty-three.

DOSTOYEVSKY, *The Devils*, 1871

Ennui, emptiness, is the constant enemy that circumvents the ageing
man in every sort of way, and meanwhile he is inevitably deprived of
the pleasantest way of spending his time: his senses no longer enable
him to enjoy simple pleasures: his legs have ceased to bear him, his
eyes to see. He must be sparing of these innocent enjoyments: even
conversation is denied him, for with whom can one speak of things
that really interest one, when one is surrounded only by people
whose age, whose prejudices are different from one's own? And yet
one wants to go on living, and one pities those that disappear. Our
weak spirits quail before the thought of ceasing to be, of ceasing to
feel either good or evil.

DELACROIX, letter to Charles Soulier, 6 August 1855

As I see it, there is not much difference between being sixty-three
and fifty-three: whereas when I was fifty-three I felt at a staggering
distance from forty-three.

SIMONE DE BEAUVOIR, 'All Said and Done', 1972

I'm fifty-three years old and six feet four. I've had three wives, five
children and three grandchildren. I love good whiskey. I still don't
understand women, and I don't think there is any man who does.

JOHN WAYNE, 1960

54

I shall soon enter into the period which, as the most agreeable of his long life, was selected by the judgement and experience of the sage Fontenelle. His choice is approved by the eloquent historian of nature, who fixes our moral happiness to the mature season in which our passions are supposed to be calmed, our duties fulfilled, our ambition satisfied, our fame and fortune established on a solid basis.

EDWARD GIBBON, *Autobiography*, 1764–91

I am fifty-four years of age—how long my productive vigour will continue to yield something good, who can tell?

RICHARD STRAUSS, letter to Hugo von Hofmannsthal, 12 July 1918

In our relationship there is no need for many words and arguments. The passing reference to your fifty-four years does more to prompt me than any amount of persuasion.

HUGO VON HOFMANNSTHAL, letter to Richard Strauss, 1 August 1918

Leopold told me this evening that I was the subject of a dialogue between Pope Pius IX and Jules Hugo, my nephew, the brother of Leopold, who died a *camerico* of the Pope. The Pope, seeing Jules, said to him:

'Your name is Hugo, is it not?'
'Yes, Holy Father.'
'Are you a relative of Victor Hugo?'
'His nephew, Holy Father.'
'How old is he?' (This was in 1857.)
'Fifty-five.'
'Alas! He is too old to return to the Church.'

VICTOR HUGO in 1870, from *Things Seen*, 1887

As life's circle draws ever tighter: all blood towards the heart. To die in love. The limbs grow pale. Life drops from me.

What love is a man can discover only at my age: the grave of life.—For me the world lives on for as long as the brain obeys the heart—How weary I am!

WAGNER, *Diary*, 16 September 1867

Having enjoyed the pleasures of the senses for long enough, I am reserving my old age for those of the mind.

SAINT-ÉVREMOND, letter to Monsieur d'Hervant, 1670

When you at a youthful fifty-four cannot avoid often thinking of death, you cannot be astonished that at the age of eighty and a half I fret whether I shall reach the age of my father and brother or further still into my mother's age, tormented on the one hand by the conflict between the wish for rest and the dread of fresh suffering that further life brings and on the other anticipation of the pain of separation from which I am still attached.

SIGMUND FREUD to Marie Bonaparte, 6 December 1936

At past fifty, Adams solemnly and painfully learned to ride the bicycle.

HENRY ADAMS, *The Education of Henry Adams*, 1905

55

'Would you agree that a woman is in her prime for about twenty, a man for about thirty years?'

'Which twenty and which thirty?'

'A woman', I replied, 'should bear children for the state from her twentieth to her fortieth year; a man should beget them for the state from the time he passes his prime as a runner until he is fifty-five . . . If any man or woman above or below these ages takes a hand in the begetting of children for the community, we shall regard it as a sin and a crime.'

PLATO, (*c.* 427–348 BC) *Republic*

Some of my friends already told me that at fifty-five I ought to give up the fabrication of love-stories.

TROLLOPE, *Autobiography*, 1883

. . . A letter from a lady who has described me in a French newspaper—'a noble lady with a shock of white hair'—Lord, are we as
old as all that? I feel about six and a half.

> VIRGINIA WOOLF, letter to Vanessa Bell, 17 August 1937

The male soprano has un bellisima (*sic*) voce but is getting on in
years. He is fifty-five.

> MOZART, letter to his sister, 7 January 1770, when he was
> nearly fourteen

Some of those young men make great and somewhat ridiculous
efforts to stifle the contradictions they have felt rising within them or
before them, without understanding that the spark of life can flash
only between two contrary poles, and that it is larger and more
beautiful the greater the distance between them and the richer the
opposition with which each pole is charged.

> ANDRÉ GIDE, *Journal*, January 1925

But still the maiden pressed me, and so I made reply,
'I'll tell you what I think my dear, about your by-and-by;
Your figure will be ampler, and, like a buzzing hive,
Your boys and girls will tease you when you are fifty-five.

'Your hair will not be brown, dear, you'll wear a decent cap;
Maybe you'll have a grandchild a-crowing on your lap;
And through the winter evenings the easiest of chairs
Will give you greater comfort than romping on the stairs.'

> R. C. Lehmann, 'Fifty Years On', *Light and Shade and Other
> Poems*, 1909

56

As when a beauteous nymph decays
We say, she's past her dancing days;
So, poets lose their feet by time,
And can no longer dance in rhyme . . .

At fifty-six, if this be true,
Am I a poet fit for you?

Or at the age of forty-three
Are you a subject fit for me?

> SWIFT, 'Stella's Birthday', 1725

Their teenish tricks, at fifty-six
All wise folk should forgo.

> Anon., c. 1800

I ought to respect myself for my friends' sake, and my children's. It is time, at fifty-six, to begin, at least, to know oneself,—and I do know what I am *not* . . .

> CONSTABLE, letter to C. R. Leslie, 1833

How could one have wished him a happier death? He died almost unconsciously in the fullness of success, and martyrdom in so great a cause consecrates his name through all history. Such a death is the crown of a noble life.

> John Stuart Mill, on the death of LINCOLN, letter to Max Kyllman, May 1865

The general, besides, was in the prime of life—that is, fifty-six, and not a day older, which under any circumstances is the most flourishing age in a man's life, the age at which *real* life can be rightly said to begin.

> DOSTOYEVSKY, *The Idiot*, 1869

This is foolish vanity. Youth is no longer essential or even becoming. Rapidly approaching fifty-seven, health and happiness are more important than lissomeness. To be fat is bad and slovenly, unless it is beyond your control, but however slim you get you will still be the age you are and no one will be fooled, so banish this nonsense once and for all, conserve your vitality by eating enough and enjoying it.

> NOEL COWARD, on the perils of slimming

57

I look into my glass,
And view my wasting skin,
And say, 'Would God it came to pass
My heart had shrunk as thin!'

For then, I, undistrest
By hearts grown cold to me,
Could lonely wait my endless rest
With equanimity.

But Time, to make me grieve,
Part steals, lets part abide;
And shakes this fragile frame at eve
With throbbings of noontide.

THOMAS HARDY, *Wessex Poems*, 1898

Age sets its house in order, and finishes its works, which to every artist is a supreme pleasure.

EMERSON, 'Old Age', *Society and Solitude*, 1870

I have almost done with harridans, and shall soon become old enough to fall in love with girls of fourteen.

SWIFT, letter to Alexander Pope, September 1725

My only fear is that I may live too long. This would be a subject of dread to me.

THOMAS JEFFERSON, letter to Philip Mazzei, March 1801

The lunches of fifty-seven years had caused his chest to slip down to the mezzanine floor.

P. G. WODEHOUSE, *My Man Jeeves*, 1919

58

All men of whatsoever quality they be, who have done anything of excellence, or which may properly resemble excellence, ought, if they are persons of truth and honesty, to describe their life with their own hand; but they ought not to attempt so fine an enterprise till they have passed the age of forty. This duty occurs to my own mind, now that I am travelling beyond the term of fifty-eight years, and am in Florence, the city of my birth. Many untoward things can I remember, such as happens to all who live upon our earth; and from those adversities I am now more free than at any previous period of my career—nay it seems to me that I enjoy greater content of soul and health of body than ever I did in bygone years. I can also bring to my mind some pleasant goods and some inestimable evils, which, when I turn my thoughts backward, strike terror in me, and astonishment that I should have reached the age of fifty-eight, wherein, thanks be to God, I am still travelling prosperously forward.

BENVENUTO CELLINI, *Autobiography*, 1562

When I was young I was fond of the speculations which seemed to promise some insight into that hidden country (the 'country of spirits'), but observing at length that they left me in the same ignorance in which they had found me, I have for very many years ceased to read or to think concerning them, and have reposed my head on that pillow of ignorance which a benevolent Creator has made so soft for us, knowing how much we should be forced to use it.

THOMAS JEFFERSON, letter to the Revd Isaac Story, 5 December 1801

. . . he [Adams] is now fifty-eight, or will be in July . . . all the Presidents were of the same age: General Washington was about fifty-eight, and I was about fifty-eight, and Mr Jefferson, and Mr Madison, and Mr Monroe.

John Adams, ex-President of the United States, on his son, JOHN QUINCY ADAMS, who had just become President. Quoted in Emerson, 'Old Age', *Society and Solitude*, 1870

Forty years on, growing older and older,
Shorter in wind, as in memory long,
Feeble of foot, and rheumatic of shoulder,
What will it help you that once you were strong?
God gives us bases to guard or beleaguer,
Games to play out whether earnest or fun;
Fights for the fearless and goals for the eager,
Twenty and thirty and forty years on.

E. E. BOWEN (1836–1901), Harrow School Song

I am a crumbling man—a magnificent ruin, no doubt, but still a ruin—and like all ruins I look best by moonlight. Give me a sprig of ivy and an owl under my arm and Tintern Abbey would not be in it with me.

W. S. GILBERT in Leslie Baily, *Gilbert and Sullivan and their World*, 1973

From the character of this crisis it is clear that what is involved is the pain of leaving a long-familiar situation, the awareness that an era in my life has come to an end, and the recognition that I must find a new basis for my existence. Despite the rigidly set ways of my fifty-eight years, I view this necessity as spiritually beneficial and I affirm it. It is also an opportunity—more stimulating than depressing—to throw off those obligations I had assumed in the course of the years out of social considerations, out of a 'sense of responsibility' or 'vanity' or what you will, and with one wrench tear myself free from the 'snares of the world', to concentrate hereafter on my own life . . .

THOMAS MANN, *Diary*, March 1933

Time, they say, makes a man mellow. I do not believe it. Time makes a man afraid, and fear makes him conciliatory, and being conciliatory he endeavours to appear to others what they will think mellow. And with fear comes the need of affection, of some human warmth to keep away the chill of the cold universe. When I speak of fear, I do not mean merely or mainly personal fear: the fear of death or decrepitude or penury or any such merely mundane misfortune. I am thinking of a more metaphysical fear. I am thinking of the fear that enters the soul through experience of the major evils to which life is subject: the treachery of friends, the death of those whom we love, the discovery of the cruelty that lurks in the average human nature.

> BERTRAND RUSSELL, *Autobiography*, ii, 1968

Trouble was all my life when things were really bad I could always take a drink and right away they were much better. When you cant take the drink is different.

> ERNEST HEMINGWAY, letter to Archibald MacLeish, June 1957

59

You've got to be fifty-nine years old t'believe a feller is at his best at sixty.

> KIN HUBBARD, *Abe Martin's Sayings*, 1915

I remember that I advised you to go on living solely to enrage those who are paying your annuities. As far as I am concerned, it is the only pleasure I have left. When I feel an attack of indigestion coming on, I picture two or three princes as gainers by my death, take courage out of spite, and conspire against them with rhubarb and temperance.

> VOLTAIRE, letter to Madame du Deffand, April 1754

This birthday opens my sixtieth year . . . I ascend a steepening path, with a burden ever gathering weight. The Almighty seems to sustain and spare me for some purpose of His own, deeply unworthy as I know myself to be. Glory be to His name!

> GLADSTONE, December 1868, soon after first becoming Prime Minister

. . . the idea that many people have that life is a vale of tears is just as false as the idea which the great majority have, and to which youth, health and wealth incline you, that life is a place of entertainment. Life is a place of service, where one sometimes has occasion to put up with a lot that is hard, but more often to experience a great many joys.

> TOLSTOY, letter to his second son, I. L. Tolstoy, October 1887

I never expected to have, in my sixties, the happiness that passed me by in my twenties.

C. S. LEWIS, letter to Nevill Coghill, 1958

Approaching sixty you must start from the beginning and see whether you can understand another's desires.

SAUL BELLOW, *Humboldt's Gift*, 1975

60

There is a very life in our despair,
Vitality of poison,—a quick root
Which feeds these deadly branches; for it were
As nothing did we die; but Life will suit
Itself to Sorrow's most detested fruit,
Like to the apples on the Dead Sea's shore,
All ashes to the taste: Did man compute
Existence by enjoyment, and count o'er
Such hours 'gainst years of life,—say, would he name three score?

BYRON, *Childe Harold's Pilgrimage*, III, xxxiv, 1812

Between thirty and forty, one is distracted by the Five Lusts;
Between seventy and eighty, one is prey to a hundred diseases.
But from fifty to sixty one is free from all ills;
Calm and still—the heart enjoys rest . . .

PO CHÜ-I (727–846 AD), 'On Being Sixty', trans. Arthur Waley

With sixty staring me in the face, I have developed inflammation of
the sentence structure and a definite hardening of the paragraphs.

JAMES THURBER, June 1955

Whilom ther was dwellynge in Lumbardye
A worthy knyght, that born was of Pavye,
In which he lyved in greet prosperitee;
And sixty yeer a wyflees man was hee,
And folwed ay his bodily delyt
On wommen, ther as was his appetyt,
As doon thise fooles that been seculeer.

CHAUCER, *The Merchant's Tale, c.* 1387

Shall I never see a bachelor of three score again?

Benedick, in SHAKESPEARE's *Much Ado About Nothing, c.* 1600

I am perhaps setting. Like a day that has been admired as a fine one, the light of it sets down amid mists and storms. I neither regret nor fear the approach of death if it is coming. I would compound for a little pain instead of this heartless muddiness of mind.

SIR WALTER SCOTT, *Journal*, September 1831

'Tis strange, that it is not in vogue to commit hara-kiri as the Japanese do at sixty. Nature is *so* insulting in her hints and notices, does not pull you by the sleeve, but pulls out your teeth, tears off your hair in patches, steals your eyesight, twists your face into an ugly mask, in short, puts all contumelies upon you, without in the least abating your zeal to make a good appearance, and all this at the same time that she is moulding the new figures around you into wonderful beauty which of course is only making your plight worse.

EMERSON, *Journals*, 1863

I feel constantly if I were but twenty years old and not over sixty all I ever wanted to do could be done easily. One never tires of life and at the last must die of thirst with the cup at one's lip.

W. B. YEATS, letter to H. J. C. Grierson, 1926

One starts to get young at the age of sixty, and then it's too late.

PICASSO

Your odious letter was on my breakfast table on the morning of my sixtieth birthday. Ah, at least one old friend has written to congratulate me on reaching the age when I am exempt from jury service and can no longer be decently expected to carry anything. Not at all. A sharp reminder that my powers are fading and that I am a bore.

EVELYN WAUGH, letter to Nancy Mitford, 29 October 1963

The uselessness of men above sixty years of age, and the incalculable benefit it would be in commercial, political, and in professional life if, as a matter of course, men stopped work at this age . . .

SIR WILLIAM OSLER, address to Johns Hopkins University, 1905

61

Oh, to be with people over sixty
Despite their tendency to prolixty!

OGDEN NASH, 'You Can Be a Republican, I'm a Gerontocrat',
Collected Verse, 1961

I observed to Lord M. that he didn't seem at all low. 'No, I'm much
better,' he replied, 'but I'm still not well.' I entreated him to take
some good advice about his health. 'That won't do any good', he
said, 'it's age and that constant care': which alas! alas! is but too true.
'I'm nearly sixty-one', he continued, 'many men die at sixty-three,
and if they get over that, live till seventy.' I told him he mustn't talk
in that way. 'People like me grow old at once, who have been rather
young for their age.' I said he still was that. 'Still, I feel a great change
since last year,' he said.

Queen Victoria (at twenty), on LORD MELBOURNE, *Diary*, 20
January 1840

After sixty, the inclination to be alone grows into a kind of real,
natural instinct; for at that age everything combines in favour of it.
The strongest impulse—the love of women's society—has little or
no effect; it is the sexless condition of old age which lays the
foundation of a certain self-sufficiency, and that gradually absorbs all
desire for others' company.

SCHOPENHAUER, 'Counsels and Maxims', *Parerga and Paralipomena*, 1851

What shall I do with this absurdity—
O heart, O troubled heart—this caricature,
Decrepit age that has been tied to me
As to a dog's tail?

W. B. YEATS, 'The Tower', 1928

The tragedy is that we don't want anything of anybody over sixty.
We don't want work from them, we don't want love from them,
often we don't even want friendship. Cowper Powys spoke of 'the
revulsion in the faces of young people which the old sometimes
glimpse'. There is a tendency to keep slightly away from old bodies.

I think the old need touching. They have reached a stage in life when they need kissing, hugging. And nobody touches them except the doctor.

RONALD BLYTHE, interview in the *New York Times*, April 1980

There are more things that I wish to explore, and I feel more alive than I did thirty years ago, when I was half my age. Young people may have more spontaneity. We have a saying in America—the pitcher is beginning to pitch with his head instead of his arm. But a creative person, if he keeps his mind alive, can still have strength in his arm. Look at Stravinsky, Picasso, and any number of writers who did such great things in later life.

ARNOLD NEWMAN, photographer, interview in the *Sunday Times Magazine*, November 1979

Scent to me is far more evocative than sound or perhaps even sight, so that I become attracted without realizing it to the smell of a floor-polish or a detergent which one day I miss when I open my door and home seems no longer home. So in my sixties I seem able to smell the leaves and grasses of my hiding-place more certainly than I hear the dangerous footsteps on the path or see the countryman's boots pass by on the level of my eyes.

GRAHAM GREENE, *A Sort of Life*, 1971

I'm an old fogey now I'm sixty-one. But I'm liberated. I never wish to run for anything again in my life.

JOHN V. LINDSAY, former Mayor of New York, 21 April 1983

62

No one is so old that he does not think he could live another year.

CICERO, *De Senectute*, *c.* 44 BC. He lived one more year

I never wake without finding life more insignificant than it was the day before . . . My greatest misery is recollecting the scene of twenty years past, and then all on a sudden dropping into the present.

SWIFT, letter to Henry St John Bolingbroke, 5 April 1729

We are a pair of old derelicts drifting around, now, with some of our passengers gone and the sunniness of others in eclipse.

MARK TWAIN, letter to W. D. Howells, January 1898

What a remarkable man! After a life incomparably rich and active, full of excitement, passion, and pleasure, he [Liszt] returns at the age of sixty-two and plays the most difficult music with the ease and strength and freshness of youth. Not only does one listen with breathless attention to his playing; one also observes the reflection in the fine lines of his face. His head, thrown back, still suggests something of Jupiter . . . Head, eyes and sometimes even a helping hand, maintain constant communication with the orchestra and audience. Sometimes he plays from notes, at other times from memory, putting on and taking off his spectacles accordingly. For the Liszt of today it was a great accomplishment; and yet he went about it as if it were nothing, and he himself the Liszt of 1840. A darling of the Gods, indeed!

Eduard Hanslick, *Vienna's Golden Years of Music 1850–1900*

I am spending delightful afternoons in my garden, watching everything living around me. As I grow older, I feel everything departing, and I love everything with more passion.

ZOLA, letter to Alfred Burneau, 2 July 1902, the year of his death

I am hunting four or five times a week here . . . it is very fine, very exciting. Even at sixty-two, I can still go harder and further and longer than some of the others. That is, I seem to have reached the point where all I have to risk is just my bones.

WILLIAM FAULKNER, letter to Joan Williams Bowen, November 1959

I want to try a very fast sail right round the world alone . . . As the years passed this urge to circle the world lay dormant in me, like a gorse seed which will lie in the earth for fifty years until the soil is stirred to admit some air or light, and the seed suddenly burgeons.

SIR FRANCIS CHICHESTER to his wife, 1964. Quoted in Anita Leslie, *Francis Chichester*, 1975

Perhaps you don't always realize that I am as fully conscious as you can be of the great apparent difference and disproportion: you in the full tide of your glorious youth: I—according to the Calendar—on the threshold at any rate of the later stages of life. Whatever comes or goes, I am and must be the debtor: hugely overdrawn, often ashamed of what I owe; but—I can't say more to-night.

H. H. ASQUITH, letter to Venetia Stanley, 30 December 1914

63

Grand climacteric (sometimes simply *the climacteric*): the sixty-third year of life (63 = 7 × 9), supposed to be specially critical. (According to some, the eighty-first year (81 = 9 × 9) was also a grand climacteric.) The phrase appears to have been taken immediately from Spanish.

Oxford English Dictionary

It is time to be old,
To take in sail.
The god of bounds,
Who sets to seas a shore,
Came to me in his fatal rounds
And said: 'No more!
No farther shoot
Thy broad ambitious branches, and thy root.
Fancy departs: no more invent;
Contract thy firmament
To compass of a tent.'

EMERSON, *Terminus*, 1886

It appears to me that in spite of myself I have been dragged to this inevitable point where old age must be undergone. I see it there before me; I have reached it; and I should at least like so to arrange matters that I do not move on, that I do not travel farther along this path of infirmities, pains, losses of memory and disfigurement. Their attack is at hand, and I hear a voice that says, 'You must go along, whatever you may say; or if indeed you will not, then you must die', which is an extremity from which nature recoils. However, that is the fate of all who go on a little too far.

> MADAME DE SÉVIGNÉ, letter to her daughter Madame de Grignan, 30 November 1689. Quoted in Simone de Beauvoir, *Old Age*, 1972

Nothing is so ridiculous as an antique face in a juvenile drawing-room.

> HORACE WALPOLE, letter to Sir Horace Mann, 31 December 1780

The happiness of finding idleness a duty. No more opinions, no more politics, no more practical tasks.

> W. B. YEATS, letter to Olivia Shakespear, 29 March 1929

Hold fast to time! Use it! Be conscious of each day, each hour! They slip away unnoticed all too easily and swiftly.

> THOMAS MANN, September 1938, *Diaries 1918–1939*

I am conscious also that during this year I have become much older. I do not think that my deafness has seriously increased, nor are my limbs less supple than they were in 1948. But I have put on weight and lost many teeth and much hair, and perhaps vigour. I am not really conscious of a decline in my mental energy. But I have intimations that my thoughts move in the old grooves and do not push out into new grooves, and that my ways of expression are becoming stereotyped. It is not a good period of history into which to grow old.

> HAROLD NICOLSON, *Diaries*, 31 December 1949

I'm sixty-three and I guess that puts me in with the geriatrics, but if there were fifteen months in every year, I'd be only forty-three.

> JAMES THURBER in 1957

At sixty-three years of age, less a quarter, one still has plans.

> COLETTE, *Evening Star*, 1973

64

Will you still need me,
Will you still feed me,
When I'm sixty-four?

LENNON and McCARTNEY, 1967

There are not many Doctor Johnsons to set forth upon their first romantic voyage at sixty-four.

ROBERT LOUIS STEVENSON, *Virginibus Puerisque*, 1881

Boswell, with some of his troublesome kindness, has informed this family, and reminded me, that the eighteenth of September is my birthday. The return of my Birthday, if I remember it, fills me with thoughts which seem to be the general care of humanity to escape. I can now look back upon threescore and four years, in which little has been done, and little has been enjoyed, a life diversified by misery, spent part in sluggishness of penury, and part under the violence of pain, in gloomy discontent, or importunate distress. But perhaps I am better than I should have been if I had been less content . . .

SAMUEL JOHNSON, letter to Mrs. Thrale, 21 September 1773

If I did not constantly repeat to myself (and why?) that I am old, I should barely feel my age. Perhaps curiosity lures me somewhat less and dawns seem to me less surprising. To the finest sunrise I say: 'Oh, let me sleep!'

ANDRÉ GIDE, *Journal*, 1934

He has a future and I have a past, so we should be all right.

JENNIE CHURCHILL, quoted by George Moore, on her marriage to Montagu Porch in 1918. He was forty-one, three years younger than her son Winston

. . . I was almost the only antediluvian. This might well have been a matter of reproach in a time of crisis, when it was natural and popular to demand the force of young men and new ideas. I saw therefore that I should have to strive my utmost to keep pace with the generation now in power and with fresh young giants who might at any time appear. In this I relied upon knowledge as well as upon all possible zeal and mental energy.

For this purpose I had recourse to a method of life which had been forced upon me at the Admiralty in 1914 and 1915, and which I found greatly extended my daily capacity for work. I always went to bed at least for one hour as early as possible in the afternoon, and exploited to the full my happy gift of falling almost immediately into deep sleep. By this means I was able to press a day and a half's work into one. Nature had not intended mankind to work from eight in the morning until midnight without that refreshment of blessed oblivion which, even if it only lasts twenty minutes, is sufficient to renew all the vital forces. I regretted having to send myself to bed like a child every afternoon, but I was rewarded by being able to work through the night until two or even later—sometimes much later —in the morning, and begin the new day between eight and nine o'clock. This routine I observed throughout the war, and I commend it to others if and when they find it necessary for a long spell to get the last scrap out of the human structure.

SIR WINSTON CHURCHILL, before becoming Prime Minister,
The Second World War, ii, 1948

It's frightfully important for a writer to be his age, not to be younger or older than he is. One might ask, 'What should I write at the age of sixty-four', but never, 'What should I write in 1940.'

W. H. AUDEN, interview in *Paris Review*, 1972

How foolish was my hope and vain
That age would conquer sin.

CHARLES WESLEY, 'In Advancing Age', 1772

65

It certainly appears that, for a man, sixty-five marks the end of that rather dangerous period which, for want of a better name, is now called his change of life. Once he has weathered this headland, he would generally set off with the wind behind him if only he would believe that he is at the beginning of the happiest stage of his voyage here below and forbid his mind to dwell upon its end.

MAURICE GOUDEKET, *The Delights of Growing Old*, 1965

. . . the next thing most like living one's own life over again, seems to be a recollection of that life; and to make that recollection as durable as possible, the putting it down in writing. Hereby, too, I shall indulge in the inclination so natural in old men to be talking of themselves and their own past actions.

BENJAMIN FRANKLIN, *Autobiography*, 1771

In spring 1775, I was struck with a disorder in my bowels, which at first gave me no alarm, but has since, as I apprehend it, become mortal and incurable. I now reckon upon a speedy dissolution. I have suffered very little pain from my disorder; and what is more strange, have, notwithstanding the great decline of my person, never suffered a moment's abatement of my spirits: insomuch, that were I to name the period of my life which I should most choose to pass over again I might be tempted to point to this later period. I possess the same ardor as ever in study, and the same gaiety in company. I consider besides, that a man of sixty-five, by dying, cuts off only a few years of infirmities: and though I see many symptoms of my literary reputation's breaking out at last with additional lustre, I know, that I had but few years to enjoy it. It is difficult to be more detached from life than I am at present.

HUME, April 1776. He died in August. *My Own Life*, 1777

Why then send Lampreys? fie, for shame!
'Twill set a virgin's blood on flame.
This to fifteen a proper gift!
It might lend sixty-five a lift.

JOHN GAY, 'To a Young Lady, with some Lampreys', *Poems on Several Occasions*, 1720, 1731

Naturally it pleases me when you write enthusiastically, as in your last letter, about my youthfulness and activity, but when I turn towards the reality principle I know it is not true and am not astonished it is not. My capacity for interest is soon exhausted: that is to say, it turns away so willingly from the present in other directions. Something in me rebels against the compulsion to go on earning money which is never enough, and to continue with the same psychological devices that for thirty years have kept me upright in the face of my contempt of people and the detestable world. Strange secret yearnings rise in me—perhaps from my ancestral heritage —for the East and the Mediterranean and for a life of quite another kind: wishes from late childhood, never to be fulfilled, which do not conform to reality as if to hint at a loosening of one's relationship to it. Instead of which—we shall meet on the soil of sober Berlin.

SIGMUND FREUD to Sandor Ferenczi, 30 March 1922

. . . I am still expecting something exciting: drinks, animated conversation, gaiety, brilliant writing, uninhibited exchange of ideas. I have never had quite the expectation of Scott Fitzgerald's character that somewhere things were 'glimmering': I thought life had its excitements wherever I was. But it was part of the same *Zeitgeist*. Now I try to discipline myself not to be so silly in depending for really deep satisfaction on things that are transitory and superficial. I try to diet and cut down on drinking and not to look forward to sprees. I hope I am well on the way to becoming a sedate old gentleman.

EDMUND WILSON, 1960

I felt as if I had been walking with destiny, and that all my past life had been but a preparation for this hour and for this trial.

SIR WINSTON CHURCHILL, on becoming Prime Minister, 11 May 1940. *The Gathering Storm*, 1948

66

He still loves life
But O O O O how he wishes
The good Lord would take him.

W. H. AUDEN, his last poem, a haiku, composed just before he died, 1973

I believe that when one is young, it is the object, the outside world, that fills one with enthusiasm—one is carried away. Later, it comes from inside: the need to express his feelings urges the painter to choose some particular starting-point, some particular form.

BONNARD, quoted in Simone de Beauvoir, *Old Age*, 1972

At the end of three months since I last wrote anything in this book, I take my pen in hand to record my determination to bring this journal (which is no journal at all) to an end. I have long seen that it is useless to carry it on, for I am entirely out of the way of hearing anything of the slightest interest beyond what is known to all the world. I therefore close this record without any intention or expectation of renewing it, with a full consciousness of the smallness of its value or interest, and with great regret that I did not make better use of the opportunities I have had of recording something more worth reading.

> CHARLES GREVILLE, 1860, concluding the journal he had kept for over forty-two years

My sixty-sixth birthday, and I still have all my old energy and passion, the same acute sensitivity, and, people tell me, the same youthful appearance.

> SOFIA TOLSTOY, *Diaries*, 22 August 1910

There is not, I think, a single example of a great painter—or sculptor—whose work has not gained in profundity and originality as he grew older. Bellini, Michelangelo, Titian, Tintoretto, Poussin, Rembrandt, Goya, Turner, Degas, Cézanne, Monet, Matisse, Braque, all produced some of their very greatest works when they were over sixty-five. It is as though a lifetime is needed to master the medium, and only when that mastery has been achieved can an artist be simply himself, revealing the true nature of his imagination.

> JOHN BERGER, *Success and Failure of Picasso*, 1965

67

Old age is no such uncomfortable thing, if one gives oneself up to it with good grace.

> HORACE WALPOLE, letter to the Countess of Ailesbury, 7 November 1774

> The pleasures that once were heaven
> Look silly at sixty-seven.
>
> NOEL COWARD, 'What's Going to Happen to the Tots?'

The evening of life I choose to pass in a quiet retreat. Ambitious projects, intrigues and quarrels of statesmen are things I have formerly been amused with, but they now seem a vain, fugitive dream. If you thought as I do, we should have more of your company, and you less of the gout.

BISHOP BERKELEY, letter to Dean Gervais, April 1752. He died the following year

Unhappily, my occupations multiply as my years increase; and yet it is almost as though, while my intellectual powers decline, my taste and urge for work grow stronger. Oh God, how much remains to be done in this glorious art, even by a man such as I have been! True, the world pays me many compliments every day, even seeing fieriness in my latest works; but no one will believe what toil and effort it costs me to produce them, for many are the days when my weak memory and the slackening of my nerves bring me so low that I sink into the most melancholy state, and am thus for many days thereafter in no condition to hit upon a single idea, until at length, Providence again having put fresh courage into my heart, I can sit down at the keyboard and begin to hammer away. And then it all comes back, thanks be to God . . .

HAYDN to his publishers, 12 June 1799

I see less well and my eyes become tired more quickly. I hear likewise less well. I tell myself that it is probably not bad that there should withdraw from us progressively an earth that one would have too much trouble leaving all at once. The wonderful thing would be, at the same time, to get progressively nearer to . . . something else.

ANDRÉ GIDE, Journal, 1937

I love being old, and can't wait to be seventy.

JOHN COWPER POWYS, letter to Nicholas Ross, 9 October 1939

The day after tomorrow I shall be sixty-seven. O lovely world, teach others to expound you as I have not been able to do!

E. M. FORSTER, Journal, 1945

68

As life runs on, the road grows strange
With faces new, and near the end
The milestones into headstones change,
'Neath every one a friend.

JAMES RUSSELL LOWELL, 'Sixty-eighth Birthday', 1889

One is past being lucky at our age.

LOUIS XIV to Marshal Villeroi after the Battle of Ramillies, 23 May, 1706

I value myself upon this, that there is nothing of the old man in my conversation. I am now sixty-eight, and I have no more of it than at twenty-eight.

SAMUEL JOHNSON to James Boswell, 30 April 1778

. . . I have added some original papers of my own, which whether they are equal or inferior to my other poems, an author is the most improper judge; and therefore I leave them wholly to the mercy of the reader. I will hope the best, that they will not be condemned; but if they should, I have the excuse of an old gentleman, who, mounting on horseback before some ladies, when I was present, got up somewhat heavily, but desired of the fair spectators, that they would count fourscore and eight before they judged him. By the mercy of God, I am already come within twenty years of his number, a cripple in my limbs, but what decays are in my mind, the reader must determine. I think myself as vigorous as ever in the faculties of my soul, excepting only my memory, which is not impaired to any great degree; and if I lose not more of it, I have no great reason to complain. What judgement I had, increases rather than diminishes; and thoughts, such as they are, come crowding in so fast upon me, that my only difficulty is to choose or to reject, to run them into verse, or to give them the other harmony of prose: I have so long studied and practised both, that they are grown into a habit, and become familiar to me. In short, though I may lawfully plead some part of the old gentleman's excuse, yet I will reserve it till I think I have greater need, and ask no grains of allowance for the faults of this

my present work, but those which are given of course to human
frailty . . .

> DRYDEN, Preface to the *Fables*, 1699. He died the following
> year

It is eleven years since I have seen my figure in a glass: the last
reflexion I saw there was so disagreeable, I resolved to spare myself
such mortifications for the future, and shall continue that resolution
to my life's end. To indulge all pleasing amusements, and avoid all
images that give disgust, is, in my opinion, the best method to attain
or confirm health.

> LADY MARY WORTLEY MONTAGU, letter to her daughter, 8
> October 1757

Nay, with some persons those awes and terrors of youth last for ever
and ever. I know, for instance, an old gentleman of sixty-eight, who
said to me one morning at breakfast, with a very agitated countenance, 'I dreamed last night that I was flogged by Doctor Raine.'
Fancy had carried him back five-and-fifty years in the course of that
evening. Dr Raine and his rod were just as awful to him in his heart,
then, at sixty-eight, as they had been at thirteen

> THACKERAY, *Vanity Fair*, 1847-8

Tomorrow I shall be sixty-nine, but I do not seem to care. I did not
start the affair, and I have not been consulted about it at any step. I
was born to be afraid of dying, but not of getting old. Age has many
advantages, and if old men were not so ridiculous, I should not mind
being one. But they *are* ridiculous, and they are ugly. The young do
not see this so clearly as we do, but some day they will.

> WILLIAM HOWELLS, letter to Mark Twain, February 1906

The knowledge that death is not so far away, that my mind and
emotions and vitality will soon disappear like a puff of smoke, has
the effect of making earthly affairs seem unimportant and human
beings more and more ignoble. It is harder to take human life
seriously, including one's own efforts and achievements and passions.

> EDMUND WILSON, 1963

69

Approaching, nearing, curious,
Thou dim, uncertain spectre—bringest thou life or death?
Strength, weakness, blindness, more paralysis and heavier?
Or placid skies and sun? Wilt stir the waters yet?
Or haply cut me short for good? Or leave me here as now
Dull, parrot-like and old, with crack'd voice harping, screeching?

WALT WHITMAN, 'Queries to my Seventieth Year', 1888

No fairy tales occur to me any more. It is as if I had filled out the entire world with fairy-tale radii close to one another. If I walk in the garden among the roses—well, what stories they and the snails have told me! If I see the broad leaf of the water-lily, then Thumbelina has already ended her journey on it. If I listen to the wind, then it has told me the journey about Valdemar Daae and has nothing better to tell me. In the wood, underneath the old oak trees, I am reminded that the Old Oak Tree has told me its last dream a long time ago. Thus I do not get any new, fresh impulses, and that is sad.

HANS ANDERSEN, letter to Mrs Melchior, 1874

But the main thing is that just as the Hindus, when they are getting on for sixty, retire to the forests, and every religious man wants to dedicate the last years of his life to God and not to jokes, puns, gossip and tennis, so I, who am entering my seventieth year, long with all my heart and soul for this tranquillity and solitude, and if not full accord, then at least not blatant discord between my life and my beliefs and conscience.

TOLSTOY, letter to his wife, July 1897

Now, being old, nearly seventy years, the sensations of colour, which give the light, are for me the reason for the abstractions which do not allow me to cover my canvas entirely nor to pursue the delimitation of the objects where their points of contact are fine and delicate; from which it results that my image or picture is incomplete.

CÉZANNE, letter to Émile Bernard, 23 October 1905. He died the following year

Perhaps the greatest change has taken place within myself as I steer towards my seventieth birthday. Most noticeable, apart from thinning hair, a paunch, paleness of eye, is the attitude of the young towards me. 'Will you be tired if we do that? Let me fetch it. I'll drive.' So far I have not yet noticed a deterioration in my drawing or my powers to write; certainly my reading has dwindled. But although I do not embark on a holiday with the same eagerness that I used to and do know that during these weeks I am not likely to fall in love, or make new friends, I am grateful that, either from my father or my mother, I have inherited a strong body and a system that withstands many trials. When I look at certain contemporaries, those that have survived, I am appalled to see how they have become prematurely old.

CECIL BEATON, *Diaries*, 23 August 1973

The score, then, to date is that I am deaf in the left ear, bald, subject to mysterious giddy fits, and practically cock-eyed. I suppose the moral of the whole thing is that I have simply got to realize that I am a few months off being seventy. I had been going along as if I were in the forties, eating and drinking everything I wanted to and smoking far too much. I had always looked on myself as a sort of freak whom age could not touch, which was where I made my ruddy error, because I'm really a senile wreck with about one and a half feet in the grave.

P. G. WODEHOUSE to his doctor, February 1951

Man's life is threescore years and ten, *
    Which God will surely bless;
Still, we are warned what follows then—
    Labour and heaviness—

And understand old David's grouch
    Though he (or so we're told)
Bespoke a virgin for his couch
    To shield him from the cold. . . .†

Are not all centuries, like men,
    Born hopeful too and gay,
And good for seventy years, but then
    Hope slowly seeps away?

* Psalms 90:10
† 1 Kings 1:1–15

True, a new geriatric art
    Prolongs our last adventures
When eyes grow dim, when teeth depart:
    For glasses come, and dentures—

Helps which these last three decades need
    If true to Freedom's cause:
Glasses (detecting crimes of greed)
    Teeth (implementing laws).

ROBERT GRAVES, 'The Imminent Seventies', *Collected Poems*, 1975

70

The days of our years are threescore years and ten; and if by reason of strength they be fourscore years, yet is their strength labour and sorrow; for it is soon cut off and we fly away.

Psalm 90

Tomorrow I will haul down the flag of hypocrisy,
I will devote my grey hairs to wine:
My life's span has reached seventy,
If I don't enjoy myself now, when shall I?
The Rubáiyát of OMAR KHAYYÁM, twelfth century, trans. Peter Avery and John Heath-Stubbs

The Massagetae in former times, Derbiccians, and I know not what nations besides, did stifle their old men after seventy years, to free them from those grievances incident to that age.
ROBERT BURTON, *The Anatomy of Melancholy*, 1621

We had a judge in Massachusetts who at sixty proposed to resign, alleging that he perceived a certain decay in his faculties; he was dissuaded by his friends, on account of the public convenience at that time. At seventy it was hinted to him that it was time to retire; but he now replied that he thought his judgement as robust and all his faculties as good as ever they were.
EMERSON, 'Old Age', *Society and Solitude*, 1870

In the seventies I was bearing in my breast,
        Penned tight,
Certain starry thoughts that threw a magic light
On the worktimes and the soundless hours of rest
In the seventies; aye, I bore them in my breast
        Penned tight.

In the seventies when my neighbours—even my friend—
    Saw me pass,
Heads were shaken, and I heard the words, 'Alas,
For his onward years and name unless he mend!'
In the seventies, when my neighbours and my friend
    Saw me pass.

In the seventies those who met me did not know
    Of the vision
That immuned me from the chillings of misprision
And the damps that choked my goings to and fro
In the seventies; yea, those nodders did not know
    Of the vision.

In the seventies nought could darken or destroy it,
    Locked in me,
Though as delicate as lamp-worm's lucency;
Neither mist nor murk could weaken or alloy it
In the seventies!—could not darken or destroy it,
    Locked in me.

THOMAS HARDY, 'In the Seventies', *Moments of Vision*, 1917

You have my acute sympathy over what you delicately call the 'nuisance' of growing old. A train has to stop at some station or other. I only wish it wasn't such an ugly and lonesome place, don't you?

RUDYARD KIPLING to a friend, on his seventieth birthday, 1935

If I had any decency, I'd be dead. Most of my friends are.

DOROTHY PARKER, 1963. Quoted in *The Algonquin Wits*, ed. Robert E. Drennan, 1968

This is the year of my seventieth birthday, a fact that bewilders me. I find it hard to believe. I understand now the look of affront I often saw in my father's face after this age and that I see in the face of my contemporaries. We are affronted because, whatever we may feel, time has turned us into curiosities in some secondhand shop. We are haunted by the suspicion that the prayers we did not know we were making have been all too blatantly answered.

V. S. PRITCHETT, *Midnight Oil*, 1971

If I had known when I was twenty-one, that I should be as happy as I am now, I should have been sincerely shocked. They promised me wormwood and the funeral raven.

> CHRISTOPHER ISHERWOOD, letter to Edward Upward, August 1974

> Seventy is wormwood,
> Seventy is gall,
> But it's better to be seventy
> Than not alive at all.

> PHYLLIS McGINLEY, on her seventieth birthday in 1975

71

A man who has settled his opinions does not love to have the tranquillity of his convictions disturbed; and at seventy-one it is time to be in earnest.

> Samuel Johnson on HECTOR MACLEAN, the minister of Coll, in *A Journey to the Western Islands*, 1773

> You think it horrible that Lust and Rage
> Should dance attendance upon my old age.
> They were not such a plague when I was young.
> What else have I to spur me into song?

> W. B. YEATS to Dorothy Wellesley, December 1936

In July, when I bury my nose in a hazel bush, I feel fifteen years old again. It's good! It smells of love!

> COROT (still feeling a compulsion to go into the country to paint), 1867

'If I have to choose between walking and painting, I'd much rather paint.' He sat down, and he never got up again. From the moment he made this important decision, it was a display of fireworks to the end. Although his palette became more and more austere, the most dazzling colours, the most daring contrasts, issued from it. It was as if all Renoir's love of the beauty of this life, which he could no longer endure physically, had gushed out of his whole tortured being. He was radiant, in the true sense of the word, by which I mean that we felt there were rays emanating from his brush, as it caressed the canvas.

> Jean Renoir on his father, AUGUSTE RENOIR, in 1912

When I was younger I could remember anything, whether it happened or not, but I am getting old and soon I shall remember only the latter.

> MARK TWAIN, letter to A. B. Paine

I still remember the refrain of one of the most popular barrack ballads of that day, which proclaimed most proudly that 'Old soldiers never die; they just fade away.' And like the old soldier of that ballad, I now close my military career and just fade away.

> GENERAL DOUGLAS MACARTHUR, address to US Senate and Congress, 19 April 1951, after having been dismissed by President Truman and before becoming chairman of Remington Rand

> You fade, old presences, and leave me here
>   In dismal trickle of a dimming May;
> I play old records and lay solitaire
>   Through aimless hours of Memorial Day.

> Cities I'll never visit, books that I'll never read,
>   Magic I'll never master. In a cage,
> I stalk from room to room, lose heat and speed,
>   Now entering the dark defile of old age.

> EDMUND WILSON, 22 May 1966

Train your will to concentrate on a limited objective. When young you spread your effort over too many things . . . If your try fails what does that matter—all life is a failure in the end. The thing is to get sport out of trying.

> SIR FRANCIS CHICHESTER, after sailing around the world, 1966–7

72

Dear Miss Martineau,

I am seventy-two years of age, at which period there comes over one a shameful love of ease and repose, common to dogs, horses, clergymen and even to *Edinburgh Reviewers*. Then an idea comes across me sometimes that I am entitled to five or six years of quiet before I die.

SYDNEY SMITH, 11 December 1842

One virtue he had in perfection, which was prudence, too often the only one that is left us at seventy-two.

GOLDSMITH, *The Vicar of Wakefield*, 1761–2

Nothing is more incumbent on the old, than to know when they shall get out of the way, and relinquish to younger successors the honours they can no longer earn, and the duties they can no longer perform.

THOMAS JEFFERSON, letter to John Vaughan, 1815

Why should not old men be mad?
Some have known a likely lad
That had a sound fly-fisher's wrist
Turn to a drunken journalist;
A girl that knew all Dante once
Live to bear children to a dunce;
A Helen of social welfare dream,
Climb on a wagonette to scream.

W. B. YEATS, 'On the Boiler', 1938

How well old people come to know that peculiar look of suppressed disgust which their obstinate concentration on some restricted sensual pleasure excites in the feverish idealism of the young and in the impatient pragmatism of the middle-aged!

JOHN COWPER POWYS, *The Art of Growing Old*, 1944

73

It is difficult for a man of seventy-three to please the public . . .

> HAYDN to the publisher Thomson in London, 1805

I'm a bit tired, but I am well. I've completely finished *Otello*!

> VERDI, letter to Opprandino Arrivabene, 1886

In age we feel again that love of our native place and our early friends, which in the bustle or amusements of middle life were overborne or suspended.

> SAMUEL JOHNSON, 1782, from Boswell's *Life*

Feeling in herself no difference between the spirits of twenty-three and seventy-three, she [Madame du Deffand] thinks there is no impediment to doing whatever one will, but the want of eyesight. If she had that I am persuaded no consideration would prevent her making me a visit at Strawberry Hill. She makes songs, sings them, remembers all that ever were made; and, having lived from the most agreeable to the most reasoning age, has all that was amiable in the last, all that is sensible in this, without the vanity of the former, or the pedant impertinence of the latter.

> HORACE WALPOLE, letter to George Montagu, 7 September 1769

Power? It has come to me too late. There were days when, on waking, I felt I could move dynasties and governments; but that has passed away.

> DISRAELI, when Prime Minister at the Congress of Berlin, June 1878

I hear the newspapers say I am dying. The charge is not true. I would not do such a thing at my time of life. I am behaving as good as I can. Merry Christmas to everybody!

> MARK TWAIN, December 1909

Old age has its compensations. I feel that I can loaf in the mornings, be less anxious about what I am going to write and not suffer afterwards so much about gaffes and errors I have made. My regrets mostly nowadays are about the things that I can't any longer do; but I dwell on old love affairs, and this does not impose upon me any further responsibility for them.

EDMUND WILSON, 1968

In the growing dark he tried to count how much time he had really lived. His brain could not cope with the simple calculation any more; three months, three weeks, a total of six months, six by eight, eighty-four . . . forty-eight thousand . . . $\sqrt{840,000}$. He summed up. 'I'm seventy-three years old, and all in all I may have lived, really lived, a total of two . . . three at the most.' And the pains, the boredom, how long had they been? Useless to try and make himself count those; the whole of the rest; seventy years.

GIUSEPPE DE LAMPEDUSA, *The Leopard*, 1958

Death? At seventy-three, I don't suppose a day goes by when it doesn't cross my mind. One hesitates to start a new book . . . It's not death that worries me, it's the process. I hope it'll come quickly, say in an air crash.

GRAHAM GREENE, interview in *The Observer*, 1978

74

Here goes a man of seventy-four,
Who sees not what life means for him,
And here another in years a score
Who reads its very figure and trim.

The one who shall walk to-day with me
Is not the youth who gazes far,
But the breezy sire who cannot see
What Earth's ingrained conditions are.

THOMAS HARDY, 'Seventy-four and Twenty', *Satires of Circumstance, 1912–13*

I have ever been esteemed one of Fortune's chiefest favourites; nor will I complain or find fault with the course my life has taken. Yet, truly, there has been nothing but toil and care and I may say that, in all my seventy-five years, I have never had a month of genuine comfort. It has been the perpetual rolling of a stone, which I have always had to raise anew . . .

GOETHE to Johann Eckermann, 1824

I have now completed my seventy-fourth year; and by the peculiar favour of God, I find my health and strength, and all my faculties of body and mind, just the same as they were at four-and-twenty.

JOHN WESLEY, *Journals*, 28 June 1777

I think Longfellow shows his added years very plainly. I went a year ago or so with him to be photographed, and the picture showed less life than any I had seen of him. This may have been temporary, but I own that he appears to me more languid in his air and movements —it is not strange at seventy-four. But I have often noticed that there are unexplained movements in health at different ages, especially in later years, both downward and upward, towards Avernus and back again, so that one who seems to be failing will grow younger again next year, and begin quite fresh after his episode of depression.

Oliver Wendell Holmes, letter to J. R. Lowell, 25 July, 1881

Without being too impolite, I should like to take leave of myself. I have decidedly seen enough of myself. I no longer even know whether or not I should still like to begin my life over again; or else I should do so with a little more daring in affirmation. I have sought much too much to please others, greatly sinned through modesty.

ANDRÉ GIDE, *Journal*, 1944

75

Since I was six, I had a mania for drawing the outlines of things. By the time I was fifty, I had published an infinite number of drawings, but all that I had produced before I was seventy was not worth counting. At seventy-five I have understood better the structure of nature, of animals, plants, trees, birds, fishes and insects. Consequently, at the age of eighty, I should have made more progress; at ninety, I should penetrate the mystery of things; at a hundred I should have reached a remarkable stage; and at 110 everything I do, every point and line, would be as a living a thing. I ask those who shall live as long to see if I keep my word. Written at the age of seventy-five years by me, formerly Hokusai, now Gawkio Rojin, the old man, mad about drawing.

HOKUSAI in 1835. He died in 1849

There can be a rewarding relationship between the sevens and the seventy-fives. They are both closer to the world of mythology and magic than all the busier people between those ages.

J. B. PRIESTLEY, 'Growing Old', *New Statesman*, July 1966

Sir, I look upon every day to be lost in which I do not make a new acquaintance.

SAMUEL JOHNSON, November 1784, from Boswell's *Life*

[Goethe] is only the building in which there once flourished a very splendid thing, and it was only that that interested me in him . . .

Heine, letter to Moses Moser, 1 July 1825

Within a few minutes I had developed, though very much more rapidly, in the same fashion as those who, after finding it hard to believe that somebody they knew in their youth had reached the age of sixty, are very much more surprised fifteen years later to learn that the same person is still alive and is only seventy-five.

MARCEL PROUST, *Time Regained*, 1927

I am getting to an age when I can only enjoy the last sport left. It is called hunting for your spectacles.

LORD GREY OF FALLODEN, in *The Observer*, 1927

Feeble old people who on such occasions learn to their surprise how highly their young contemporaries esteem them are often overcome by their excess of emotion, and a little later succumb to the after-effects. You get nothing for nothing, and you have to pay heavily for living too long!

SIGMUND FREUD, on his birthday celebrations, 1931

I am ready to meet my Maker. Whether my Maker is prepared for the ordeal of meeting me is another matter.

SIR WINSTON CHURCHILL, on his seventy-fifth birthday, 1949

Reith decided that he would celebrate his seventy-fifth birthday by climbing, for the last time, to the top of Cairngorm . . . He told a friend that he had one great hope and longing in pursuing this apparently foolhardy enterprise: 'It's only a dream but I've a feeling I may meet my father again at the summit. I'd like to die up there and go away with him.'

Andrew Boyle on LORD REITH in 1964

And now every fresh day finds me more filled with wonder and better qualified to draw the last drop of delight from it. For up until now I had never known time's inexpressible wealth; and my youth had never entirely yielded itself to happiness. Is it indeed this that they call growing old, this continual surge of memories that come breaking in on my inner silence, this contained and sober joy, this lighthearted music that bears me up, this spreading kindly feeling and this gentleness?

MAURICE GOUDEKET, *The Delights of Growing Old*, 1965

I have learned to read the papers calmly and not to hate the fools I read about.

EDMUND WILSON, 1970

Raymond (Mortimer) said that he had been happier during the last five years (he is seventy-five) than at any other time of his life.

Cecil Beaton, *Diaries*, 11 September 1973

I have always admired the Esquimaux. One fine day a delicious meal is cooked for dear old mother, and then she goes walking away over the ice, *and doesn't come back* . . .

One should be proud of leaving life like that—with dignity and resolution.

AGATHA CHRISTIE, *Autobiography*, 1977

76

If I go on eating with as good an appetite as I do at present, I shall be the ruin of a whole host of Englishmen who have wagered huge sums that I will die before the 1st of September.

LOUIS XIV, during supper on 18 June 1715. He did, in fact, die on 1 September

I was a hard student until I entered on the business of life, the duties of which leave no idle time to those disposed to fulfil them; and now, retired, and at the age of seventy-six, I am again a hard student.

THOMAS JEFFERSON, letter to Dr Vine Utley, 21 March 1819

I approach the term when my daily journal must cease from physical disability to keep it up. I have now struggled nearly five years, without the interval of a day, while mind and body have been wearing away under the daily, silent, but unremitting erosion of time. I rose this morning at four, and with smarting, bloodshot eye, and shivering hand, still sat down and wrote to fill up the chasm of the closing days of last week, but my stern chase after Time is, to borrow a simile from Tom Paine, like the race of a man with a wooden leg after a horse.

JOHN QUINCY ADAMS, 25 March 1844

In the course of a conversation with C. G. Jung in 1951, when he was seventy-six, I asked him if he would give me a message for the elderly for whom I was helping to care. He said: 'Tell them to live each day as if they'll be here for another hundred years. Then they will really live to the end.'

Barbara Robb, letter to the Daily Telegraph, July 1973

One of the delights of being older is being able to control ideas. I have suffered all my life from a disease called Brains in the Head . . . In youth you keep bubbling with ideas. They may be foolish but you can't stop them. I've now learnt not to suffer too much from the Brains . . . As you get older your judgement develops. One of my joys is having my mind stirred by a good book, and not feeling I've got to go to the typewriter afterwards. There is nothing nicer than nodding off while reading. Going fast asleep then being woken up by the crash of the book on the floor, then saying to yourself, well it doesn't matter much. An admirable feeling.

A. J. P. TAYLOR, interview in Evening Standard, March 1982

I have only managed to live so long by carrying no hatreds.

SIR WINSTON CHURCHILL to Lady Violet Bonham-Carter, 15 October 1951

77

Nowadays everybody wants to be young,
so much so, that even the young are old with the effort of being
    young.
As for those over fifty, either they rush forward in self-assertion
fearful to behold,
or they bear everybody a grim and grisly grudge
because of their own fifty or sixty or seventy or eighty summers.
As if it's my fault that the old girl is seventy-seven!

> D. H. LAWRENCE, 'Old People', *Pansies*, 1929

I can hardly think I am entered this day into the seventy-eighth year
of my age. By the blessing of God, I am just the same as when I
entered the twenty-eighth. This hath God wrought, chiefly by my
constant exercise, my rising early, and preaching morning and
evening.

> JOHN WESLEY, *Journals*, 28 June, 1780

I do not like drowsiness—mine is an old man's natural infirmity, and
that same old man creeps upon me more and more. I cannot walk
him away: he gets hold on the memory, and my poor little accounts
never come right. Let me nevertheless be thankful: I have very little
pain . . . We are all in health, for I will not call my lassitude and
stupidity by the name of illness. Like Lear, I am a poor old man and
foolish, but happily I have no daughter who vexes me.

> GEORGE CRABBE, shortly before his death in 1832, described
> by his son in 1834

If Providence gives me another two years, I believe I shall still be able
to paint something beautiful.

> COROT, 1873

Oh, would that my mind could let fall its dead ideas, as the tree does
its withered leaves! And without too many regrets, if possible! Those
from which the sap has withdrawn. But, good Lord, what beautiful
colours!

> ANDRÉ GIDE, *Journal*, 1947

My father proceeded to dismiss the guests with a little ceremonial flutter of the hand, and went on to tell me at once of his serious illness, although there were no outward signs of it in his physical appearance. I therefore asked to be allowed to talk to his doctor, Giglioli . . . When I asked him if my father was very ill, he replied:

'The disease which Sir George assumes he has developed is a matter of X-ray plates and not of faith alone. He is an old man of seventy-seven years of age, and of course if he gets a cold it may always turn to pneumonia: but probably he will outlive me and be with us for another twenty years.' (My father did in fact survive his doctor by some five.)

On my return from the city I found the invalid still in bed, and looking intensely depressed. As I opened the door of his room he enquired in a drooping voice:

'How long does the doctor give me?'

'About twenty years,' I replied.

At this he suddenly jumped out of bed with an agility that would have done credit to a man half his age, and said: 'I must dress now.' That night he came down for dinner.

> SIR GEORGE SITWELL in Osbert Sitwell, *Tales My Father Taught Me*, 1979

I'm an old man now and so, though I don't wish you to make any allowances, you must be kind. No, I've no notion of giving up. That's really in God's hands. At the same time I should add that if the devil or the angels desert me, then I'll call it a day.

> VLADIMIR HOROWITZ, before giving his first London piano recital for thirty-one years, interview in *The Observer*, 1982

78

I still exist, and still enjoy some pleasure in that existence, though now in my seventy-ninth year. Yet I feel the infirmities of age come on so fast, and the building to need so many repairs, that in a little time the owner will find it cheaper to pull it down and build a new one.

> BENJAMIN FRANKLIN, letter to Mrs Mary Hewson, 19 March 1784

He studies ten hours every day; writes constantly without spectacles, and walks out with only a domestic, very often a mile or two.

VOLTAIRE's servant, talking about his master to Charles Burney

The only 'funny' thing I remember his saying, was, on one occasion when we were accidentally left alone in the dark, after some jesting remark on the danger of my reputation—'Ah, my dear, if sweet, seventy-eight could come again! *Mais ces beaux jours sont passés.*'

Helen Selina, Lady Dufferin, letter to Abraham Hayward, 8 February 1858, recalling the poet SAMUEL ROGERS

In my young days, sickly as I was, I could spend ten or even twelve hours at my desk!! working all the time; and more than once I worked from four in the morning till four in the afternoon with no more than a cup of coffee . . . and working without pausing for breath. But now I can't do it. In those days I was in control of my own physique and of time itself . . . but now, alas, I can't do it . . . So to conclude: The best thing will be to tell everybody, from now on, that I can't and will not make the slightest promise about *Falstaff.* If it comes off, well and good; and we shall see how it turns out!

VERDI, letter to Giulio Ricordi, January 1891

Don't complain about old age. How much good it has brought me that was unexpected and beautiful. I concluded from that that the end of old age and of life will be just as unexpectedly beautiful . . .

TOLSTOY, letter to V. V. Stasov, September 1906

'I have lived seventy-eight years without hearing of bloody places like Cambodia.'
    With a whimsical look he strung out half a dozen strange-sounding names.
    'They have never worried me, and I haven't worried them.'

SIR WINSTON CHURCHILL to Lord Moran, 1953

I've never really learnt how to live, and I've discovered too late that life is for living.

LORD REITH in an interview with Malcolm Muggeridge, 1967

79

. . . when everything takes longer to do—bathing, shaving, getting dressed or undressed—but when time passes quickly, as if he were gathering speed while coasting downhill. The year from seventy-nine to eighty is like a week when he was a boy.

MALCOLM COWLEY, *The View from 80*, 1980

Lord Palmerston after the division scrambled up a wearying staircase to the ladies' gallery. My informant, who was behind him, had the good taste and tact to linger. He saw the ladies' gallery open and Lady Palmerston advanced and they embraced! An interesting scene and what pluck—to mount those dreadful stairs at three o'clock in the morning at eighty years of age! My informant would not disturb them. It was a great moment! But still Lady de Grey who with other Whig Ladies was in the gallery with Lady Palmerston would come forward with 'O! dear Lord Palmerston! How nice!' etc. etc. which spoilt all.

Disraeli's Reminiscences, 1864

Although in his eightieth year, Verdi had spent hours each day at the rehearsals [of *Falstaff*] and then more hours back at his hotel dealing with all the admirers, friends, critics, bores and telegrams. Strepponi marvelled at his resilience, as did all the journalists who wrote articles on everything from his clothes to his digestion . . . The rehearsals exhilarated him, and he evidently had determined to go through the ordeal of the traditional first three performances with the best grace he could. He took endless curtain calls, appeared on the balcony of his hotel and answered hundreds of congratulatory notes and telegrams. Only one rumour agitated him into action. A newspaper declared that the government intended to give him the title of 'Marquis of Busseto'. Verdi telegraphed at once and in horror to the Minister of Education, who replied that the rumour was baseless.

VERDI in 1892. George Martin, *Verdi: His Music, Life and Times*, 1965

Here I am, already on the eve of the fourth year of my pontificate, with an immense programme of work in front of me to be carried out before the eyes of the whole world, which is watching and waiting.

POPE JOHN XXIII, August 1961

During each of the last ten years I have written more poems and discarded fewer than at any other time, with the exception of 1974 when I wrote only five. I notice, however, that the annual survival rate of new poems, prior to 1965, was in fact five, and now in my eightieth year it is certain to rise significantly again.

ROBERT GRAVES, Introduction to *Collected Poems*, 1975

I constantly hear the Goddess of Fate whispering in my ear: You won't last much longer . . .

ANDRÉ GIDE, *Journal*, 8 June 1948

80

Pray, do not mock me:
I am a very foolish fond old man,
Fourscore and upward, not an hour more, or
    less:
And, to deal plainly,
I fear I am not in my perfect mind.

SHAKESPEARE, *King Lear*, c. 1608

In a dream you are never eighty.

ANNE SEXTON, 'Old', *Selected Poems*, 1962

One should never be sorry one has attempted something new
—never, never, never.

SYBIL THORNDIKE, after the flop of *Vanity Fair*, a musical in
which she danced and sang at the Queen's Theatre, London, in
November 1962

At eighty I believe I am a far more cheerful person than I was at
twenty or thirty. I most definitely would not want to be a teenager
again. Youth may be glorious, but it is also painful to endure.
Moreover, what is called youth is not youth; it is rather something
like premature old age.

HENRY MILLER, 'On Turning Eighty', *Sextet*, 1977

Eighty years old! No eyes left, no ears, no teeth, no legs, no wind!
And when all is said and done, how astonishingly well one does
without them!

PAUL CLAUDEL, *Journal*, 1948

An artist's early work may generally be said to be his most serious
rival, but in the Taureaux de Bordeaux, Goya, at the age of eighty
odd years, actually surpassed himself. Movement takes the place of

form—the tremulous excitement of the crowds of spectators watching the sweaty drama in the ring, the rush to and fro of the *torreros*, the stubborn strength of the short powerful brute with the man impaled on his horns, the dust and glitter and riot of the scene is rendered in a most extraordinary manner.

Goya died in 1828, at the age of eighty-two. W. Rothenstein, GOYA, 1900

In February 1865 the Speaker of the House of Commons dined with the Prime Minister, Lord Palmerston, then eighty. The Speaker told Disraeli that Palmerston 'ate for dinner two plates of turtle soup; he was then served very amply to a plate of cod and oyster sauce; he then took a pate; afterwards he was helped to two very greasy-looking entrees; he then despatched a plate of roast mutton; there then appeared before him the largest, and to my mind the hardest, slice of ham that ever figured on the table of a nobleman, yet it disappeared, just in time for him to answer the inquiry of his butler: "Snipe, my Lord, or pheasant?" He instantly replied pheasant, thus completing his ninth dish of meat at that meal.'

Disraeli's Reminiscences, 1865

Alas! I am gone very far down the vale of years, a vale in which there is no fine prospect on the other side, and the few flowers are scarcely worth gathering. But it is pleasant to turn round in the midst of one's weariness and look on the verdant declivity behind; pleasant to see pure white images on either hand, and to distinguish here and there a capital letter on the plinth.

WALTER SAVAGE LANDOR, letter to Rose Paynter, 18 January 1856

All my life I have hated anniversaries of every sort. It's a ridiculous habit. At this advanced age, when there is nothing left to think about but death, they want to bother me with that!

TOLSTOY, *Diary*, 20 September 1909

What is the secret meaning of this celebrating the big round numbers of one's life? Surely a measure of triumph over the transitoriness of life, which, as we never forget, is ready to devour us. Then one rejoices with a sort of communal feeling that we are not made of such frail stuff as to prevent one of us victoriously resisting the hostile effects of life for sixty, seventy or even eighty years. That one can understand and agree with, but the celebration evidently has sense only when the survivor can in spite of all wounds and scars join in as a hale fellow; it loses this sense when he is an invalid with whom there is no question of conviviality. And since the latter is my case and I bear my fate by myself, I should prefer my eightieth birthday to be treated as my private affair—by my friends.

> SIGMUND FREUD to Ernest Jones, 1936

I was aware at the time that some people noted a certain discrepancy in our ages—a bridegroom is not usually thirty years *older* than his father-in-law.

> PABLO CASALS, cellist, talking about his marriage to Marta Montanez y Martinez in 1957

But why, you ask me, should this tale be told
To men grown old, or who are growing old?
It is too late! Ah, nothing is too late
Till the tired heart shall cease to palpitate.
Cato learned Greek at eighty; Sophocles
Wrote his grand *Oedipus*, and Simonides
Bore off the prize of verse from his compeers,
When each had numbered more than fourscore years.
And Theophrastes, at fourscore and ten
Had but begun his *Characters of Men*.
Chaucer, at Woodstock with the nightingales,
At sixty wrote the *Canterbury Tales*;
Goethe at Weimar, toiling to the last,
Completed *Faust* when eighty years were past.
These are indeed exceptions, but they show
How far the gulf-stream of our youth may flow
Into the arctic regions of our lives,
Where little else than life itself survives.

> LONGFELLOW, 'Poem for the Fiftieth Anniversary of the Class of 1825 in Bowdoin College'

81

I believe that if Christ had not been crucified, and had lived out the term of His life as allotted by nature, that He would have been translated into His eighty-first year from a mortal to an eternal body.
> DANTE, *The Banquet*, 1304–8. Eighty-one, or 9 × 9, was considered the perfect age by mystics. See also p. 122

> Resigned to live, prepared to die,
> With not one sin but poetry,
> This day Tom's fair account has run
> (Without a blot) to eighty-one.
>> POPE, 'Tom Southerne's Birth-day Dinner at Ld. Orrery's', 1742

Well! at eighty-two years old, and my eighty-first birthday is hard at hand, one is easily convinced of money's importance to felicity. *No* suicide, or comparatively none, is committed but for *lack of pelf.* Yet money, if people are *stuffed* with it, like a fillet of veal, does not keep them alive.
> MRS PIOZZI, letter to Mrs Pennington, 27 December 1820

My remaining days I may now consider a free gift; and it is now, in fact, of little consequence what I now do, or whether I do anything.
> GOETHE to Johann Eckermann, after completing *Faust*, June 1831

Again my old birthday returns, my eighty-first! God has been very merciful and supported me, but my trials and anxieties have been manifold and I feel tired and upset by all I have gone through this winter and spring.
> QUEEN VICTORIA, *Diary*, on her last birthday, 24 May 1900

When one has reached eighty-one one likes to sit back and let the world turn by itself, without trying to push it.
> SEAN O'CASEY, *New York Times*, September 1960

Time flies swifter and ever swifter. I dare say if I live long enough, perhaps another twenty years, time may cease to exist for me. The seasons seem so short. The longest day is so quickly followed by the equinoctial one and that by the shortest. As for the moons, I scarcely measure their flight any more. The single days slip between my fingers. Perhaps because I live fairly in the forenoon alone. The rest of the day I exist merely, unless some very stimulating person rouses me out of invading torpor.

BERNARD BERENSON, *Diary*, 1947

There was an old man who said 'Damn;
    What the hell of a nuisance I am.
      The girls are so kind,
      They don't seem to mind
    But—What the *hell* of a nuisance I am.'

A. P. HERBERT to Lady Davidson, 1972

The last concert of my career was the one I gave at the Wigmore Hall in London for the benefit of the hall, which was in danger of being demolished. My concert was to give an example to other artists in order to save the old endangered place. As for myself, it was a symbolic gesture; it was in this hall that I had given my first recital in London and playing there for the last time in my life made me think of my whole career in the form of a sonata. The first movement represented the struggles of my youth, the following andante for the beginning of a more serious aspect of my talent, a scherzo represented well the unexpected great success, and the finale turned out to be a wonderful moving end.

ARTUR RUBINSTEIN, *My Many Years*, 1980

If anything I am too young for the part. Puccini wrote it to be sung *con voce stanca da vecchio decrepito*—with an old croaky voice. I said to the conductor: 'Do you want me to sing in my normal voice, or as I'll sing in ten years time when I am ninety?'

HUGUES CUENOD, at eighty-one, before singing the Emperor in Puccini's *Turandot* in London, November 1982

I like my work, so why should I retire?

LORD DENNING, Master of the Rolls, 23 January 1980. Lord Denning became a judge in 1944, before Parliament fixed the retiring age of judges at seventy-five. He eventually retired in 1982

82

Sophocles, Plato, Socrates,
   Gentlemen,
Pythagoras, Thucydides,
Herodotus, and Homer,—yea,
Clement, Augustin, Origen,
Burnt brightlier towards their setting-day,
   Gentlemen.

And ye, red-lipped and smooth-browed; list,
   Gentlemen;
Much is there waits you we have missed;
Much love we leave you worth the knowing,
Much, much has lain outside our ken:
Nay, rush not: time serves: we are going,
   Gentlemen.

THOMAS HARDY, from 'An Ancient to Ancients', *Late Lyrics and Earlier*, 1922

I beg you not to say that I am only eighty-two: it is a cruel calumny. Even if it be true, according to an accursed baptismal record, that I was born in November 1694, you must always agree with me that I am in my eighty-third year.

VOLTAIRE, letter to Argantal, 1777

I am delighted to find that even at my great age ideas come to me the pursuit and development of which would require another lifetime.

GOETHE

I wish that it could occasionally occur to your Reflection, that I am eighty-two not twenty-eight years of Age . . .

Speak gently to the aged one
Grieve not the care-worn heart;
The sands of life are nearly run;
Let such in peace depart.

DUKE OF WELLINGTON to Angela Burdett-Coutts (aged thirty-seven), 1851

When the dumb Hour, clothed in black,
Brings the Dreams about my bed,
Call me not so often back,
Silent Voices of the dead,
Toward the lowland ways behind me,
And the sunlight that is gone!
Call me rather, silent voices,
Forward to the starry track
Glimmering up the heights beyond me
On, and always on!

TENNYSON, 'The Silent Voices', 1892, shortly before his death

Now, as I approach my eighty-fourth year—it seems even older
when I see it in print—I find it interesting to reflect on what has made
my life, even with its moments of pain, such an essentially happy
one. I have come to the conclusion that the most important element
in human life is faith. If God were to take away all His blessings:
health, physical fitness, wealth, intelligence and leave me but one
gift, I would ask for faith—for with faith in Him, in His goodness,
mercy, love for me and belief in everlasting life, I believe I could
suffer the loss of my other gifts, etc, and still be happy—trustful,
leaving all to His Inscrutable Providence.

ROSE FITZGERALD KENNEDY, *Times to Remember*, 1974

I am rather prone to senile lechery just now—want to touch the right
person in the right place, in order to shake off bodily loneliness.

E. M. FORSTER, letter to Joe Ackerley, 16 October 1961

There is never a moment when a painter can say, 'I've done a good
day's work and tomorrow is Sunday' . . . You can never write the
words The End.

PICASSO

Well, you're tempting me. You know, Mr Gladstone formed his last
government when he was eighty-three, and I'm only eighty-two
—you mustn't put temptation in my way.

HAROLD MACMILLAN in 1976 when asked by Robin Day how
he would react if called on to form a coalition government

You couldn't live eighty-two years in the world without being disillusioned.

REBECCA WEST in *The Observer*, 1975

83

. . . For a man who is always living in the midst of these studies and labours does not perceive when old age creeps upon him. Thus, by slow and imperceptible degrees life draws to its end. There is no sudden breakage; it just goes slowly out.

CATO, speaking in Cicero's *De Senectute*, *c.* 44 BC

My present life is no dead-and-alive existence, but most active, and of such a kind that all people esteem me fortunate. I am always in good health, and am so nimble that without help I can leap lightly upon a horse, and can without effort climb not only a flight of stairs, but a steep acclivity.

LUIGI CORNARO, *Of the Moderate Life and the Art of attaining a Great Age*, 1553

I am a wonder to myself! I am never tired (such is the goodness of God) either with writing, preaching or travelling. One natural cause, undoubtedly, is my continual exercise, and change of air. How the latter contributes to health, I know not; but certainly it does.

JOHN WESLEY, *Journal*, 1786

Eighty-three years gone by! I do not know that I am satisfied when I consider how so many years have passed, how I have filled them. What useless agitations, what fruitless endeavours! Tiresome complications, exaggerated emotions, spent efforts, wasted gifts, hatreds aroused, sense of proportion lost, illusions destroyed, tastes exhausted! What results in the end? Moral and physical weariness, complete discouragement and profound disgust with the past. There are a crowd of people who have the gift or the drawback of never properly understanding themselves. I possess only too much the opposite disadvantage or superiority; it increases with the gravity of old age.

> TALLEYRAND, 2 February 1837. He retired at the age of eighty
> from being ambassador to London

I represent the youth and hope of England. The solution of these questions of the future belongs aright to us who are of the future, and not to you who are of the past.

> GLADSTONE, speech in the House of Commons, 1893

Poor Gladstone! How he has aged: he positively looks older than Randolph Churchill and so very undistinguished: and how senile! I could not hear a word he said, except when occasionally he cheered Asquith in a quavering, common voice; he reminded me of a figure in a picture of Michaelmas Day in the Almshouses.

> Max Beerbohm, letter to Reggie Turner, 15 April 1893

I understood now why it was that the Duc de Guermantes, who to my surprise, when I had seen him sitting on a chair, had seemed to me so little aged although he had so many more years beneath him than I had, had presently, when he rose to his feet and tried to stand firm upon them, swayed backwards and forwards upon legs as tottery as those of some old archbishop with nothing solid about his person but his metal crucifix, to whose support there rushes a mob of sturdy young seminarists, and had advanced with difficulty, trembling like a leaf, upon the almost unmanageable summit of his eighty-three years, as though men spend their lives perched upon living stilts which never cease to grow until sometimes they become taller than church steeples, making it in the end both difficult and perilous for them to walk and raising them to an eminence from which suddenly they fall.

> MARCEL PROUST, *Time Regained*, 1927

For the first time I have lately become aware of the fact that the period of our earthly existence is limited. During the whole of my life this idea has never actually come into my mind. It occurred to me very distinctly when I was looking at an old tree there in the garden. When we came it was very small, and I looked at it from above. Now it waves high above my head and seems to say 'You will soon depart, but I shall stay here for hundreds more years.'

> SIBELIUS, 1948

84

In former courts polite and gay,
And still a beauty in decay.
Go mourn in form, yet shed no tears,
Such falls give life to happy heirs,
Who can lament a full ripe death,
When eighty-five resigns its breath;
So plums in autumn's fruitful crop,
Mellowed by time, corrupt and drop.

Anon., 'Epitaph on the Duchess of Marlborough', 1744

When we are young, we are ignorant of life, and the beguilements of its movements, its distractions, its excesses fascinate us, so that we accept the little ups and downs without being aware of living. Now we are conscious of life, we feel it, and its sorrow oppresses and crushes us.

VERDI

I, personally, have succeeded in living nearly eighty-five years without taking any trouble about my diet.

BERTRAND RUSSELL, 25 March 1957

Why do I get stuck down in the past? Why do I keep going over and over those years when I know I cannot change anything?

SIR WINSTON CHURCHILL to Lord Moran, 1959

'So good to see you looking so well', I said and he [Picasso] ran to touch wood. He does not like the process of getting old. 'Come back again in twenty years, but come back sooner as well.'

Cecil Beaton, *Diaries*, 28 April 1965

My eighty-fourth birthday. Should be feeling solemn, but there is little solemnity in me and that little never about myself. The past year has brought decided diminution of energy. I get tired more and more quickly. I have to forget walks of forty minutes at the utmost.

Forty minutes is the limit of profitable concentration whether on active reading, writing or enjoyment of concentration.

BERNARD BERENSON, *Diary*, 26 June 1949

But the nicest thing about this second childhood is the link it brings with the first childhood! When a year ago Phyllis and I were sitting with a circle of grown-ups in the Doctor's Waiting-Room waiting our turn to be called into the surgery, there was a tiny toddler, too little for me to know whether it was a baby-boy or a baby-girl. But after a few minutes of surveying each other, this tiny tot waved its hand to me, and I waved back! It was just as if we had said to each other: 'Lord! What fools these grown-ups be!'

JOHN COWPER POWYS, letter to Nicholas Ross, 1957

85

Forty and forty-five are bad enough; fifty is simply hell to face; fifteen minutes after that you are sixty; and then in ten minutes more you are eighty-five.

DON MARQUIS

The harvest is gathered in. At the age of eighty-five one has the right, perhaps the duty, of falling silent.

SAINT-SAËNS, 1920

How do you do? As we have been friends for seventy years, and are candidates for promotion to another world, where I hope we shall be better acquainted, I think we ought to inquire now and then after each other's health and welfare, while we stay here. I am not tormented with the fear of death, nor, though suffering under many infirmities, and agitated by many afflictions, weary of life. I have a better opinion of this world and its Ruler than some people seem to have. A kind Providence has preserved and supported me for eighty-five years and seven months, through many dangers and difficulties, though in great weakness, and I am not afraid to trust in its goodness to all eternity . . .

JOHN ADAMS, letter to David Sewell, 22 May 1821

My supreme thought on my eighty-fifth birthday is one of inexpressible gratitude for the opportunities which life has brought me of being of service to my fellow men.

J. D. ROCKEFELLER, July 1924

To me, old age is always fifteen years older than I am.

BERNARD M. BARUCH, 20 August 1955

Old age is only half as funny as one is inclined to think . . . The older I grow the more impressed I am by the frailty and uncertainty of our understanding, and all the more I take recourse to the simplicity of immediate experience so as not to lose contact with the essentials, namely the dominants which rule human existence throughout the milleniums . . .

C. G. JUNG to Lord Sandwich, who congratulated him on his eighty-fifth birthday, 10 August 1960

I am Joan Miró . . . but I work every day and I want to die shouting *mierda*.

JOAN MIRÓ, 21 April 1978

86

I now find I grow old. My sight is decayed, so that I cannot read a small print, unless in a strong light. My strength is decayed, so that I walk much slower than I did some years since. My memory of names, whether of persons or places, is decayed, till I stop a little to recollect them. What I should be afraid of is, if I took thought for the morrow, that my body should weigh down my mind, and create either stubbornness, by the decrease of my understanding, or peevishness, by the increase of bodily infirmities.

JOHN WESLEY, on his eighty-sixth birthday, 28 June 1789. 'His strength now diminished so much that he found it difficult to preach more than twice a day.' Robert Southey, *The Life of John Wesley*, 1820

Well, World, you have kept faith with me,
　　Kept faith with me;

Upon the whole you have proved to be
      Much as you said you were.
Since as a child I used to lie
Upon the leaze and watch the sky,
Never, I own, expected I
      That life would all be fair . . .

THOMAS HARDY, 'A Consideration on my eighty-sixth birthday', *Winter Words*, pub. 1928

I found him just going out coursing with his servants, three
greyhounds and three spaniels; he wished to turn back but I preferred
accompanying him. We beat the common fields two hours. I
happened to find a hare setting, which the dogs killed after a long
course, which the old gentleman seemed highly interested in. How
surprising it is to see a man between eighty and ninety years of age
(really eighty-six) enjoying the strength and spirits to enter into
amusements of this kind; he has had the living of Bennington
fifty-four years, and probably in the whole of that time has not
missed a season either coursing or shooting. I see no manner of
alteration in him, and he is exactly the same as when I used to visit
him before I went to Oxford, which is nearly twenty years ago.

John Skinner describing a visit to his uncle, the REVD JOHN
HAGGARD, Rector of Bennington, Herts, in 1810

The speech at Liverpool which he delivered at the age of eighty-six,
lasted for an hour and twenty minutes. It was the last great public
speech of his life, and it evoked round upon round of applause.
Gladstone denounced the Sultan of Turkey as 'the great assassin' . . .

GLADSTONE in 1896

The journey across the stage to the piano was negotiated with
faltering care, and the piano-stool, once achieved, was comman-
deered with the triumphant air of a captain investing a besieged
garrison. Yet when the nimble fingers skimmed over the keyboard
and the architecture of the music began to appear in all its clear
majestic lines, his hearers were reassured that three-quarters of a
century spent in playing to the world's audiences had failed to blunt
the sharpness of his style or the keenness of his attack.
    He celebrated his eighty-sixth birthday in October by conducting
rehearsals for the long-forgotten Ascanio at the Opera, where he
supervised the work of the chorus in draughty little rooms and

crouched in the pit to solve difficult orchestral passages with the players.

SAINT-SAËNS in 1921, the year of his death

When I returned to my country home at Houjarray on the evening of May 9, 1975, freed from all outside responsibilities for the first time for many years, there lying on my table was the first sketch-plan of this book, a new and exciting task for which I was very little prepared.

JEAN MONNET, *Memoirs*, 1978

I shall be going to church as usual. Then we shall have a family lunch and I shall have my rest. You can hardly expect me to go out ski-ing or anything at my age.

HAROLD MACMILLAN, interview in *The Observer*, 10 February 1980

My dear, when you are my age you will realize that what you need is the maturer man.

LADY DIANA COOPER, when her name was coupled with that of Sir Robert Mayer, soon to be a hundred years old

87

My long and tangled life this day concludes its eighty-seventh year. My father died four days short of that term. I know no other life so long in the Gladstone family, and my profession has been that of politician, or, more strictly, minister of state, an extremely short-lived race when their scene of action is the House of Commons, Lord Palmerston being the only complete exception . . . Still old age is appointed for the gradual loosing and succeeding snapping of the threads . . .

GLADSTONE at the age of eighty-seven. *Diary*, 29 December 1896. His last entry after seventy years

So far as I am aware, I happen to be the only English poet who has brought out a new volume of verse on his . . . birthday, whatever may have been the case with the ancient Greeks, for it must be remembered that poets did not die young in those days . . .

This being probably my last appearance on the literary stage, I would say, more seriously, that though, alas, it would be idle to pretend that the publication of these poems can have much interest for me, the track having been adventured so many times before today, the pieces themselves have been prepared with reasonable care, if not quite with the zest of a young man new to print.

> THOMAS HARDY, Introductory Note to *Winter Words*, his last collection of poetry, published posthumously in 1928 with the birthday left blank

At the end of the 'Bacchanale' he stepped down from the podium and prepared to leave the stage: but Miller (the first cellist) stopped him and reminded him that he still had to conduct the *Meistersinger* Prelude. Toscanini nodded, returned to the podium and beat numbly through the last piece. As the orchestra hammered out those final affirmative bars of C major, as the strings tumbled downwards towards the final resolution, the numbness must have given way to a feeling of suffocation, to a need for release and escape stronger than his ironclad sense of duty towards the music and towards the orchestra; for while the last chords were still being played, he stopped his baton, stepped slowly from the podium and made his way to the exit.

> TOSCANINI's last concert at Carnegie Hall, 4 April 1954

I agree with you that being thrown to the wolves is tedious, but after eighty-seven years of it my reactions are no longer so vehement as they once were.

> BERTRAND RUSSELL to a correspondent, 1960

I find now at eighty-seven that I forget people's names occasionally and more regrettably, owing to my wretched eyesight, people's faces, but mercifully my power to concentrate on work in here has not in the least diminished.

> COMPTON MACKENZIE, *Octave Ten*, 1971

In March 1978 Picasso embarked on a series of etchings in which, using much the same cast as in the drawings, he stages an outright Saturnalia . . . The etchings liberate the paintings. Now the painter may embrace his model: not with chastity, since for Picasso art was never chaste, nor with any sense of shame—since 'only when painting isn't really painting can there be an affront to modesty' —but with carnality and love.

Denis Thomas, PICASSO *and His Art*, 1975

'I have always lived an unhealthy life', he said. When he was eighty-seven I noticed for the first time that he did not any more have a small brandy at lunch. A letter in which he was questioned about the influence of tobacco and alcohol on his work went unanswered, but he said to me: 'All the doctors who wanted to forbid me to smoke and drink are dead. But I am quietly going on living. It's not every man of my age who can unhesitatingly eat and drink as I do.'

SIBELIUS, 1972

I think I have lived long enough and worked hard enough, not without some recognized achievement, to have earned the privilege of not being addressed as 'Pussycat'!

STRAVINSKY, 1970

88

One thing I have certainly learnt—ńot to have a good opinion of myself.

DEAN INGE, 1948

No leisure. That is what I miss most at my age. I have collected the books that I looked forward to reading at ease when I was old. I planted a garden where I could stroll with a book of verse open in my hands, and occasionally looking at the trees and flowers. Now that I am old I have no time for either, least of all for browsing among my books. This diary is full of these complaints. Perhaps the greatest disappointment of all is the lack of spacious days with nothing to accomplish but the freedom to enjoy the harvest of one's years. Quite the contrary! Attention to health takes up most of the time that is not absorbed by long rests, sleeps, dozings. The hours when I can read are so reduced that I scarcely dare embark on a serious book, and

I waste myself on periodicals, snippets and light 'literature'. Woe is me!

BERNARD BERENSON, *Diary*, 11 April 1954

I asked [Bernard Shaw] how he was occupying his time, and he told me that he had been studying the newest books on Pitman's short-hand: 'I wished to refresh myself by learning the latest changes. Some days I find it difficult to write—my right hand wobbles—and I want to send my articles to the papers in shorthand.' His speed of authorship, he said, still averaged about fifteen hundred words a day, and he continued to give interviews, write articles, review books, and send letters to *The Times* on military, religious, political, medical and economic subjects.

Hesketh Pearson, BERNARD SHAW, 1961

There was the time when my cousin Harold congratulated me on the condition of our London garden: I demurred and said quite truthful-ly that my father, then aged eighty-eight, should take a lot of credit. 'Oh, no,' said my father without a trace of irony, 'I only do the heavy work.'

Hallam Tennyson recalling his father SIR CHARLES TENNY-SON in *The Listener*, 1968

I was overwhelmed by flattery. But to use the words of the Press: 'booksey success was not likely to spoil me'.

The REVD WILLIAM KEBLE MARTIN, after being acclaimed as Author of the Year for 1965, having published his first book *The Concise British Flora* after sixty years working on it

89

Monsieur Daron, very perplexed, became excited. 'But he certainly died of something! What is your opinion?'

The doctor raised his arms. 'I absolutely do not know. He died because he died, that's all.'

Then Monsieur Daron, in a voice full of emotion, demanded: 'Exactly how old was that one? I can't remember.'

'Eighty-nine.'

And the little old man, with an air at once incredulous and reassured, cried, 'Eighty-nine! So it wasn't old age! . . .'

GUY DE MAUPASSANT, An Old Man, 1882

The old President sat in a large stuffed arm-chair, dressed in a blue coat, black small-clothes, white stockings; a cotton cap covered his bald head. We made our compliment, told him he must let us join our congratulations to those of the nation of the happiness of his house. He thanked us, and said: 'I am rejoiced, because the nation is happy. The time of gratulation and congratulations is nearly over with me; I am astonished that I have lived to see and know of this event. I have lived now nearly a century; a long, harassed, and distracted life.'

Emerson describing a visit to ex-President JOHN ADAMS in
February 1825, shortly after the election of his son John Quincy
Adams as President. 'Old Age', Society and Solitude, 1870

I am nearly blind and totally deaf. My son Charles undresses me, and I do not give any trouble. I dine on soup . . .

WALTER SAVAGE LANDOR to Robert Browning, 22 August
1864. He died the following month

Comparing the world in which I live now with the one into which I was born, I might just as well have been born three hundred years, instead of ninety years, ago. Scarcely anything we now consider as indispensable to normal living existed in the world of my birth. It was a world without automobiles or aeroplanes, without radio, television or movies, without miracle drugs, electric home appliances, fountain pens, or frozen foods . . .

BERNARD M. BARUCH, The Public Years, 1960

What in heaven's name have I got to look forward to? I'm very old—in my ninetieth year. I have a horrible dislike of old age. Everybody's dead—half, no nearly all of one's contemporaries —and those that aren't are ga-ga. Someone rang the other day and said 'I want to invite you and Duff over for dinner.' I said, 'But Duff's been dead for twenty-eight years.' (taps forehead). That's what I dread.

LADY DIANA COOPER, interview in Sunday Times, September
1981

The year 1976 during which I reached my ninetieth birthday was by far the most eventful and rewarding of my lifetime, for two of my farces were revived and I found myself the rather bewildered old author of three simultaneous London successes (*The Bed, Plunder* and *Banana Ridge*).

BEN TRAVERS, *A-Sitting on a Gate*, 1978

The change of the seasons is today the most important thing in my life.

SIBELIUS, 1955

90

Tho' I am now past ninety, and too old
T' expect preferment in the Court of Cupid,
And many winters made me ev'n so cold
I am become almost all over stupid.

Yet I can love and have a mistress too,
As fair as can be and as wise as fair;
And yet not proud, nor anything will do
To make me of her favour to despair.

To tell you who she is were very bold;
But if i' th' character your self you find
Think not the man a fool tho' he be old
Who loves in body fair, a fairer mind.

THOMAS HOBBES, 1679, the year of his death

At ninety they [the Struldbruggs] lose their teeth and hair, they have
at that age no distinction of taste, but eat and drink whatever they can
get, without relish or appetite. The diseases they were subject to still
continue without increasing or diminishing. In talking they forget
the common appellation of things, and the names of persons, even of
those who are their nearest friends and relations. For the same reason
they never can amuse themselves with reading, because their memory will not serve to carry them from the beginning of a sentence to
the end; and by this defect they are deprived of the only entertainment whereof they might otherwise be capable.

SWIFT, *Gulliver's Travels*, 1726

It was characteristic that a few months before his death [Sophocles]
appeared with his chorus and actors in mourning for Euripides at the
proagon [dress-parade] before the Great Dionysia.

SOPHOCLES in 406 BC. *Oxford Classical Dictionary*, 1949

Charles Macklin, the Shakespearian actor, made his last appearance on the stage in May 1789, when he dressed for the part of Shylock, his favourite role. 'Seeing Mrs Pope he asked her if she was playing that night. She answered that she dressed for Portia. "Ah, very true," said Macklin, "but who is to play Shylock?" He went on to the stage, spoke a few lines of his part, then making an apology, he quitted the stage for ever.'

CHARLES MACKLIN, described by Joseph Knight, *Dictionary of National Biography*, 1893

The riders in a race do not stop short when they reach the goal. There is a little finishing canter before coming to a standstill. There is time to hear the kind voice of friends and to say to oneself: 'The work is done.' But just as one says that, the answer comes: 'The race is over, but the work never is done while the power to work remains.' The canter that brings you to a standstill need not be only coming to rest. It cannot be, while you still live. For to live is to function. That is all there is in living.

OLIVER WENDELL HOLMES Jr., 8 March 1931, in a radio address as a Supreme Court judge on his ninetieth birthday

The way in which the world has developed during the last fifty years has brought about in me changes opposite to those which are supposed to be typical of old age. One is frequently assured by men who have no doubt of their own wisdom that old age should bring serenity and a larger vision in which seeming evils are viewed as means to an ultimate good.

I cannot accept any such view. Serenity, in the present world, can only be achieved through blindness or brutality. Unlike what I conventionally expected, I became gradually more and more of a rebel. I was not born rebellious. Until 1914, I fitted more or less comfortably into the world as I found it. There were evils—great evils—but there was reason to think that they would grow less. Without having the temperament of a rebel, the course of events has made me gradually less and less able to acquiesce patiently in what is happening. A minority, though a growing one, feels as I do, and so long as I live, it is with them that I must work.

BERTRAND RUSSELL, 13 May 1962

The next morning, when in the bright sunlight I started on the first sitting, Adenauer had, as his secretary told me, already spent two hours working on his memoirs, either dictating, which he mostly did standing up, or leafing through reference books from the shelves. I had a comfortable armchair placed on the rostrum, so that he should not overtire himself. To stand and pose for two or three hours every day—and he was, after all, over ninety—would be too much for him, I thought. He rebuked me with a grin; after all, he said, I did not sit down to paint, and we both belonged to a generation that does not age.

ADENAUER in Oskar Kokoschka, *My Life*, 1974

My old lady died of a common cold.
She smoked cigars and was ninety years old.
She was thin as paper with the ribs of a kite,
And she flew out the kitchen door one night.

Now I'm no younger'n the old lady was,
When she lost gravitation, and I smoke cigars.
I feel sort of peaked, an' I look kinda pore,
So for God's sake, lock that kitchen door!

TENNESSEE WILLIAMS, 'Kitchen Door Blues', 1946

91

Confirmed also that Thomas Hobbes died at Hardwick within twelve miles of Chatsworth, that on his deathbed he should say that he was ninety-one years finding out a hole to go out of this world, and at length he found it. He died on 4 Dec. Thursday.

Anthony à Wood on HOBBES, 1679

A man over ninety is a great comfort to all his elderly neighbours: he is a picket-guard at the extreme outpost; and the young folks of sixty and seventy feel that the enemy must get by him before he can come near their camp.

OLIVER WENDELL HOLMES, *The Guardian Angel*, 1867

Every day in my old age is more important than I can say. It will never return. When one takes one's leave of life one notices how much one has left undone.

SIBELIUS, 1957, shortly before his death

I commit this to print within a few weeks of completing my ninety-second year [i.e. at the age of ninety-two]. At such an age I should apologise for perpetrating another play or presuming to pontificate in any fashion. I can hardly walk through my garden without a tumble or two; and it seems out of all reason to believe that a man who cannot do a simple thing like that can practise the craft of Shakespear. Is it not a serious sign of dotage to talk about oneself, which is precisely what I am now doing? Should it not warn me that my bolt is shot, and my place silent in the chimney corner?

Well, I grant all this; yet I cannot hold my tongue nor my pen. As long as I live I must write. If I stopped writing I should die for want of something to do . . .

BERNARD SHAW, Preface to *Far-fetched Fables*, 1948

My nonagenarian conclusion about people is that essentially they will act according to their nature because they must. They may be constrained, they may be obliged to seem to conform to the values entertained by the umpires of the game. In hidden ways they will go their way, and only *force majeure* will stop them. I have seldom if ever encountered a person whose fundamental nature I could see modified. Wherefore one should avoid individuals whose ego is invincibly opposed to ours. No matter what their intentions may be, they act up to their own nature. To a sensitive person they ooze and smell and look what they are—I mean really that we all do—and we should avoid those that at first sight repel us. My worst troubles have come from trying to conquer instinctive apathies, alarm signals.

BERNARD BERENSON, *Diary*, 16 September 1956

My whole attitude to life is spiritual—a feeling of identification with all nature, all mankind, all life, the whole of the past, the whole of the future.

LORD (FENNER) BROCKWAY, interview in *New York Times*, 1979

At age ninety-one Frank Lloyd Wright 'seemed a permanent part of the American scene, as much a fixture as an historic landmark. He was rarely sick, always active, vigorous, and very conspicuous. As March turned into April 1959, he accepted another commission, granted another interview, continued work on the Marin County Civic Centre and several other projects, and fired off an indignant letter denouncing Monona Terrace opponents: everything business as usual.'

Robert C. Twombly, FRANK LLOYD WRIGHT, 1979

I did affirm to my readers in *My Young Years* that I was the happiest man I had ever met and I can profoundly reaffirm it at the age of ninety-one.

ARTUR RUBINSTEIN, *My Many Years*, 1980

You know that I'm at death's door. But the trouble is that I'm afraid to knock.

SOMERSET MAUGHAM to his nephew Robin, 1965

92

What I wouldn't give to be seventy again!

OLIVER WENDELL HOLMES Jr., after seeing a pretty girl

When you are ninety-two and you say, 'When I was seventy-four', it is almost like saying, 'When I was young!'

A clergyman's widow, quoted in Ronald Blythe's *The View in Winter*, 1979

'Mr President, give me leave to ask you a question I have sometimes asked of aged persons, but never of any so aged or so learned as yourself.' He looked so kindly at me that I thought I might go on. 'Every studious man, in the course of a long and thoughtful life, has had occasion to experience the special value of some axiom or precept. Would you mind giving me the benefit of such a word of advice?' He bade me explain, evidently to gain time. I quoted an instance. He nodded and looked thoughtful. Presently he brightened up and said, 'I think, sir, since you care for the advice of an old man, sir, you will find it a very good practice' (here he looked me in the face) *'always to verify your references*, sir!'

John William Burgon interviewing DR MARTIN ROUTH, President of Magdalen College, Oxford, 1847. *Quarterly Review*, July 1878

I start my ninety-third year with no fear of dying, but dread of getting blinder and deafer, cut off from the outer world, from reading and seeing with my own eyes—the greatest joys of the past, and reduced to what surges up from the depths of memory. What does surge up is seldom recollection of happiness, of good doing, of what others owe to me, but of the meannesses, pettinesses, nastinesses, dishonest actions committed by me in the past, even the distant ones of my childhood.

BERNARD BERENSON, *Diary*, 27 June 1957

There seems to be something about England that keeps its writers alive to a ripe old age, and we are thinking of settling down here. One of my English friends is going to take me to meet Walter de la Mare, who is eighty-two, but I won't be able to meet Max Beerbohm, eighty-three, since he is in Italy . . . Eden Phillpotts is now writing for television at the age of ninety-two, and today's *Times* notes that H. M. Tomlinson celebrated his eighty-second birthday yesterday, quietly at home. In America, as you know, most male writers fail to reach the age of sixty, or, if they do, they have nothing more to say, but occasionally say it anyway.

JAMES THURBER, letter to Ronald and Jane Williams, 23 June 1955

Oh Lord, how old *old age* IS. One doesn't (one *cannot*) conceive how old and awful old age is, till one is up to the eyes. Then . . . one says puff and blows it all out of the open empty window.

GORDON CRAIG to his son Edward, 1964

The trouble about reaching the age of ninety-two, which I did last October, is that regrets for a misspent life are bound to creep in, and whenever you see me with a furrowed brow you can be sure that what is on my mind is the thought that if only I had taken up golf earlier and devoted my whole time to it instead of fooling about writing stories and things, I might have got my handicap down to under eighteen.

P. G. WODEHOUSE, Preface to *The Golf Omnibus*, 1973

93

In the memory of GEORGE NEWTON,
of Stalybridge,
who died August 7th, 1871,
in the 94th year of his age.

Though he liv'd long, the old man has gone at last,
No more he'll hear the huntsman's stirring blast;
Though fleet as Reynard in his youthful prime,
At last he's yielded to the hand of Time.

Blithe as a lark, dress'd in his coat of green,
With hounds and horn the old man was seen.
But ah! Death came, worn out and full of years,
He died in peace, mourn'd by his offspring's tears.

Inscription in the graveyard of Mottram, Yorkshire

For the past eighty years I have started each day in the same manner.
It is not a mechanical routine but something essential to my daily life.
I go to the piano, and I play two preludes and fugues of Bach. I
cannot think of doing otherwise. It is a sort of benediction on the
house. But that is not its only meaning for me. It is a rediscovery of
the world of which I have the joy of being a part. It fills me with
awareness of the wonder of life, with a feeling of the incredible
marvel of being a human being.

PABLO CASALS in *Joys and Sorrows*

I get up awfully early, have my breakfast about quarter past eight,
and then I generally work after breakfast. When I've got any work to
do, when I'm doing a novel, I always give up the morning to it. And
then, after lunch, Ethel and I go out to pick up the mail, take the two
dogs down to the water, run them around, and get back here about
five o'clock. After that I just read.

P. G. WODEHOUSE interviewed by the BBC at his home on
Long Island, soon after the publication of his last book, *Aunts
Aren't Gentlemen*, 1975

But I'll tell you the really great thing about living to be ninety-three: one does not have any rivals, because they're all dead, so one can afford to be generous with young chaps like you.

BEN TRAVERS to Alan Ayckbourn, *The Times*, November 1979

At every stage of my life friendship has been the main source of my quite outrageously enjoyable existence. My sorrows, even though occasionally acute, have been few, and the enjoyment still goes on. A great many of my friends, both within and outside my family circle, are sixty or seventy years younger than myself.

SIR GEOFFREY KEYNES, *The Gates of Memory*, 1981, concluded, 'soon after entering my ninety-fourth year'. He died in 1982

94

Nay, they [my humours] are still perfect: nor is it possible they should be otherwise in my present condition, when I find myself hearty and content, eating with a good appetite, and sleeping soundly. Moreover, all my senses are as good as ever, and in the highest perfection; my understanding clearer and brighter than ever; my judgement sound; my memory tenacious; my spirits good; and my voice, the first thing which is apt to fail us, grown so strong and sonorous, that I cannot help chanting out loud my prayers morning and night, instead of whispering and muttering them to myself, as was formerly my custom. . . .

O, how glorious this life of mine is like to be, replete with all the felicities which man can enjoy at this side of the grave; and even exempt from that sensual brutality which age has enabled my better reason to banish; because where reason resides, there is no room for sensuality, nor for its bitter fruits, the passions and perturbations of the mind, with a train of disagreeable apprehensions. Nor yet can the thoughts of death find room in my mind, as I have no sensuality to nourish such thoughts. Neither can the death of grandchildren and other relations and friends make any impression on me, but for a moment or two; and then it is over.

LUIGI CORNARO, *An Earnest Exhortation to a Sober Life*, 1560, see also p. 184

It was interesting . . . to see coming back through the aged features of your father the lineaments of the very handsome young man one sees in his early photographs.

C. Walter Hodges describing GORDON CRAIG, 1966

I don't want to see ANYBODY, and I don't want anybody to see me. You don't know what it is to be old as I am. Do you suppose I want the great GBS to be remembered as a doddering old skeleton?

BERNARD SHAW to Hesketh Pearson, June 1950

95

A woman of ninety said to M. de Fontenelle, when he was ninety-five: 'Death has forgotten us.' 'Sssh!' de Fontenelle answered, putting his finger over his mouth.

Quoted by Nicholas Chamfort, *Characters and Anecdotes 1741–94*

Right in front of him—probably because he was stone-deaf, and it was deemed more edifying to hear nothing at a short distance than at a long one—sat 'Old Maxum', as he was called, his real patronymic remaining a mystery to most persons. A fine philological sense discerns in this cognomen an indication that the pauper patriarch had once been considered pithy and sententious in his speech; but now the weight of ninety-five years lay heavy on his tongue as well as his ears, and he sat before the clergyman with protruded chin, and munching mouth, and eyes that seemed to look at emptiness.

GEORGE ELIOT, *Scenes from Clerical Life*, 1858

Now that I am ninety-five years old, looking back over the years, I have seen many changes take place, so many inventions have been made, things now go faster, in olden times things were not so rushed. I think people were more content, more satisfied with life than they are today, you don't hear nearly as much laughter and shouting as you did in my day, and what was fun for us wouldn't be fun now . . . In this age I don't think people are as happy, they don't take time to be happy, they are worried. They're too anxious to get ahead of their neighbours, they are striving and striving to get something better. I do think in a way that they have too much now. We did with much less.

GRANDMA MOSES, 1955

So how do I live now, when in my ninety-sixth year? Until a few months ago I alternated between my typewriter and my kitchen. Now, apart from attendance at the House of Lords, a few functions —less now than ever—I am engaged in the dictation of a book, a task I was reluctant to undertake, and have only done so under pressure; more of a challenge than an adventure . . . If, when not in good form, I have to make a speech or answer a question in the Lords I feel much better. It is stimulating, even if one's speech is unacceptable or even ignored.

LORD (MANNY) SHINWELL, *Lead with the Left: My First Ninety-Six Years*, 1981

If you think about it, you will find that there is no meaning in life if you are estranged from God.

CATHERINE BRAMWELL-BOOTH, former Commissioner of the Salvation Army and granddaughter of its founder, 1979

96

I never thought that I'd survive,
That I'd contrive to stay alive
and whoop it up at ninety-five.
But, damn it all, I find that I've
    Increased the score
    To one year more.
        WOW!
And now, you know, it seems to me
That even one full century
Need not be necessarily
A real impossibility . . .

FRANK BUXTON, former editor of the *Boston Herald*, 'At Ninety-Six', 1973

[Bertrand Russell's] memory was still exceptional, and Christopher Farley remembers that in the course of conversation Russell mentioned how the existence of pi, the constant ratio between the circumference and the diameter of a circle, had been known in biblical times. But they had got it wrong, he said, and went on to quote chapter and verse. When Farley discovered that the quotation was word perfect he asked when Russell had last read that particular book of the Bible. 'Oh, thirty or forty years ago', replied the man who believed in knowing his enemies.

> Ronald W. Clark, *The Life of* BERTRAND RUSSELL, 1975

Princess Alice still makes an occasional official appearance, although she is a bit frail. But not too frail to read without eyeglasses, not too frail to make an occasional shopping sortie by bus, not too frail to attend Sunday services at St Mary Abbots Church near her house in Kensington. When friends pressed her to carry a walking stick, she reluctantly agreed, but she had it disguised as an umbrella.

> R. W. Apple, describing PRINCESS ALICE, Countess of Athlone and the last surviving granddaughter of Queen Victoria, in 1979

97

. . . blind old Dandolo, elected Doge at eighty-four years, storming Constantinople at ninety-four, and after the revolt again victorious and elected at the age of ninety-six to the throne of the Eastern Empire, which he declined, and died Doge at ninety-seven.

> ENRICO DANDOLO, Doge of Venice, who died in 1205. Emerson, 'Old Age', *Society and Solitude*, 1870

On Sunday morning, 23 May 1937, at 4.05 o'clock, Rockefeller died at 'The Casements'. His health had been good to the last, though he tired easily and had to be given constant attendance . . . Just before he died he was displaying a keen interest in plans for some remodelling at 'The Casements', and for an air conditioning system at Lakewood.

> Allan Nevins, JOHN D. ROCKEFELLER, 1940

I do so hate to leave this world.

> BERTRAND RUSSELL to his wife Edith, January 1970, a month
> before he died

98

The oldest people today are not substantially older than people
twenty-five centuries ago. Pythagoras, in the sixth century BC, lived
to be ninety-one. Heraclitus of Ephesus died at ninety-six, Isocrates,
an Athenian orator, at ninety-eight. The exceptional survival time of
yesterday is still the exceptional survival time of today.

> Dr Andrew Leaf, Professor at Harvard Medical School, 1978

The last period of his life was mainly occupied in writing: his intellect
remained unimpaired to the end, for at the age of ninety he published
one of his most important works, the *Philippus*; and between 342 and
339 he wrote the *Panathenaicus*. His last composition was a letter
congratulating Philip of Macedon on his victory at Chaeronea (338).
He died a few days later.

> ISOCRATES (436–338 BC). *Oxford Classical Dictionary*, 1949

Looking back on my long life I can truly say that in any line of
activity which chance has opened to me, I have tried 'to be George
Schuster'—to concentrate on this activity as an end in itself. I have
taken any chance and challenge as it came. I have never envisaged any
sort of plan for a career . . .

I see the sense of work well done as the essential element in the
foundation for true happiness in the world.

> SIR GEORGE SCHUSTER, *Private Work and Public Causes: A
> Personal Record 1881–1978*, 1979

The doctors don't want me to be active. Ha! Why, in New York,
two or three times a week, weather permitting, I jump into a cab and
go to my office . . . Infirmity is a state of mind. I sleep well because I
don't let my mind get stale.

> ADOLPH ZUKOR, Chairman Emeritus of Paramount Film Stu-
> dios, 1971

99

She drank good ale, strong punch and wine,
    And lived to the age of ninety-nine.

Epitaph to MRS FRELAND, in Edwelton churchyard, Nottinghamshire, 1741

Nor is age any bar to our maintaining pursuits of every other kind, especially of agriculture, to the very extreme verge of old age. For instance, we have it on record that M. Valerius Corvus kept it up to his hundredth year, living on his land and cultivating it after his career was over, though between his first and sixth consulships there was an interval of six and forty years. So that he had an official career lasting the number of years which our ancestors defined as coming between birth and the beginning of old age. Moreover, that last period of his old age was more blessed than that of his middle life, inasmuch as he had greater influence and less labour. For the crowning grace of old age is influence.

CICERO, *De Senectute, c.* 44 BC

I feel nothing except a certain difficulty in continuing to exist.

BERNARD DE FONTENELLE. Fontenelle died in January 1757, a month short of his hundredth birthday. He was perpetual secretary of the French Academy of Sciences for forty-two years. His last major work, *Apologie des Tourbillons*, was written in 1752, when he was ninety-five

The President of Magdalen has sufficient eyesight left by daylight to scrawl his best thanks to Mrs Whorwood for her kind note and her congratulations on his having completed his ninety-ninth year. May God who disposes everything for the best grant us to meet happily in a better world. I am . . . my dear Madam

Your Obliged Servant
M. J. Routh.

Sept. 21 1854

DR MARTIN ROUTH, theologian, died in 1854 while still President of Magdalen College, Oxford, a post he held for sixty-three years

These docs, they always ask you how you live so long. I tell 'em, 'If I'd known I was gonna live this long, I'd have taken better care of myself.'

> EUBIE BLAKE, jazz composer and pianist, interview, on his hundredth birthday, February 1983. He died a week later

I must not forget to tell you of the death of a fellow of Trinity College aged ninety-seven. His funeral was attended by a brother of ninety-nine. The latter was much distressed and said he had always told his junior that theological research was not compatible with longevity. 'God,' he solemnly told Rutherford, 'does not mean us to pry into these matters.' After the funeral the old man went back to Trinity and solemnly drank his half-bottle of port. He was asked his prescription for health, and said with great fervour, 'Never deny yourself anything.' He explained that he had never married as he found fidelity restricting as a young man. 'I was once engaged when I was forty,' he said, 'and I found it gave me very serious constipation. So I broke off the engagement and the lady quite understood.' He was very anxious not to be thought past the age of flirtation.

> *The Correspondence of Mr Justice Holmes and Harold J. Laski*, 1953

100

He was extraordinarily sober . . . and dieted himself always with so much wisdom and precaution, that finding his natural heat decaying by degrees in old age, he also diminished his diet by degrees, so far as to stint himself to the yolk of an egg for a meal, and sometimes, a little before his death, it served him for two meals. By this means he preserved his health, and was also vigorous, to the age of a hundred years.

> Letter from a nun of Padua, about her grandfather LUIGI CORNARO, who died in 1566. See also p. 177

Just before her 101st birthday I visited her and found her in good spirits. She was mentally alert and full of plans. 'As soon as I get back home, I will start painting again,' she said: 'Much as I enjoy visiting with my friends and neighbors, I have come to see that *one* hundred-year celebration is enough for anybody, and I would like to spend my 101st birthday the same as my first day, very quiet.'

> Otto Kallir, GRANDMA MOSES, 1973. She died in 1961, aged 101

I have dreamt since my youth that divine providence would grant me the privilege of dying in harness.

> SIR ROBERT MAYER, *My First Hundred Years*, 1979

Who wants to live to be a hundred? What's the point of it?

A short life and a merry one is far better than a long life sustained by fear, caution and perpetual medical surveillance.

> HENRY MILLER, 'On Turning Eighty', 1971, in *Sextet*, 1977

At my age I stand, as it were, on a high peak alone. I have no contemporaries with whom I can exchange memories or views. But that very isolation gives me a less biased view of that vast panorama of human life which is spread before the eyes of a centenarian, still more when those eyes are the eyes of an archaeologist. It is true that much of the far distance is shrouded in cloud and mist, but every here and there the fog thins a little and one can see clearly the advance of mankind.

DR MARGARET MURRAY, *My First Hundred Years*, 1963. She died the same year

ACKNOWLEDGEMENTS

❧

THE editors and publisher gratefully acknowledge permission to reproduce copyright material in this book:

Kingsley Amis: 'Ode to Me' from *Collected Poems 1944–1979*. Copyright © 1972 Kingsley Amis. First printed in *The Observer*. Reprinted by permission of Hutchinson Publishing Group Ltd., and Jonathan Clowes Ltd., London, on behalf of the author; extract from *One Fat Englishman*. Reprinted by permission of Victor Gollancz Ltd., and Jonathan Clowes Ltd.

Daisy Ashford: from *The Young Visiters*. Reprinted by permission of the author's Literary Estate and Chatto & Windus.

W. H. Auden: 'Letter to Lord Byron'. Copyright 1937 by Wystan Hugh Auden. Reprinted from *W. H. Auden: Collected Poems*, ed. Edward Mendelson, by permission of Faber & Faber Ltd., and of Random House, Inc.

Peter Avery & John Heath-Stubbs: from *The Rubáiyát of Omar Khayyhám*, trans. Peter Avery and John Heath-Stubbs (Penguin Classics 1981). Copyright © Peter Avery and John Heath-Stubbs 1979. Reprinted by permission of Penguin Books Ltd.

Bernard M. Baruch: from *Baruch: The Public Years*. Copyright © 1960 by Bernard M. Baruch. Reprinted by permission of Laurence Pollinger Ltd., and of Holt, Rinehart and Winston, Publishers, Inc.

Sir Cecil Beaton: extracts from his *Diaries*. Reprinted by permission of his Literary Executor and George Weidenfeld & Nicolson Ltd.

Simone de Beauvoir: from *Old Age*, trans. Patrick O'Brien. Reprinted by permission of George Weidenfeld & Nicolson Ltd.; from *All Said and Done*, trans. Patrick O'Brien. Copyright © Editions Gallimard, 1972. English translation © André Deutsch, Weidenfeld & Nicolson, and G. P. Putnam's Sons. Reprinted by permission of George Weidenfeld & Nicolson Ltd., G. P. Putnam's Sons and Editions Gallimard; from *The Prime of Life*, trans. Peter Green. Reprinted by permission of George Weidenfeld & Nicolson Ltd., and Editions Gallimard.

Max Beerbohm: from *Max Beerbohm: Letters to Reggie Turner*, ed. Rupert Hart-Davis, © Eva Reichmann 1964; and from 'A Small Boy Seeing Giants', published in *Mainly on the Air* (Heinemann, 1946). Reprinted by permission of Mrs Eva Reichmann.

Hilaire Belloc: extracts from 'Godolphin Horne', 'Lord Lundy' and 'Sarah Byng' from *The Complete Verse*. Reprinted by permission of Gerald Duckworth & Co., Ltd. From *Napoleon*. Reprinted by permission of A. D. Peters & Co., Ltd.

Bernard Berenson: from *Sunset and Twilight* (Harcourt Brace Jovanovich, 1963: Hamish Hamilton, 1964).

John Berger: from *Success and Failure of Picasso*. Reprinted by permission of the author.

John Betjeman: extract from 'The Wykehamist' from *Collected Poems*. Reprinted by permission of John Murray (Publishers) Ltd.

Bruno Bettelheim: from *The Children of the Dream*. Copyright © 1969 by Macmillan Publishing Co., Inc. Reprinted by permission.

Andrew Boyle: from *Only the Wind Will Listen: Reith of the BBC*. Reprinted by permission of the Hutchinson Publishing Group Ltd.

Professor S. Brodetsky: from *Sir Isaac Newton* (Methuen & Co., 1927). Reprinted by permission of Associated Book Publishers Ltd.

Rupert Brooke: from a letter to Edward Marsh, April 1914, in *Rupert Brooke* by Christopher Hassell (Faber, 1964). Copyright Faber & Faber Ltd., and the Estate of Rupert Brooke. Reproduced by permission.

Pablo Casals and Albert E. Kahn: from *Joys and Sorrows. Reflections by Pablo Casals and Albert E. Kahn*. Reprinted by permission of Macdonald & Co. (Publishers) Ltd.

Cézanne: from *Letters*, ed. and trans. John Rewald. Reprinted by permission of Bruno Cassirer (Publishers) Ltd.

Sir Henry Channon: from *The Diaries of Sir Henry ('Chips') Channon*, ed. Robert Rhodes James. Reprinted by permission of George Weidenfeld & Nicolson Ltd.

Anton Chekhov: from Chekhov: *Plays*, trans. Elisaveta Fen (Penguin Classics, Revised Edition, 1954). Copyright © Elisaveta Fen, 1951, 1954. Reprinted by permission of Penguin Books Ltd.

Francis Chichester: from *Francis Chichester* by Anita Leslie. Reprinted by permission of Hodder & Stoughton Ltd., and Hutchinson Publishing Group Ltd.

Sir Winston Churchill: extracts from Lord Moran: *Churchill, The Struggle For Survival*. Reprinted by permission of Constable & Co., Ltd; extracts from *My Early Life*. Reprinted by permission of The Hamlyn Publishing Group (originally published by Odhams Press Limited) and Charles Scribner's Sons; extracts from *The Second World War, Volume 2: Their Finest Hour*. Copyright 1949 by Houghton Mifflin Company, copyright © renewed 1976 by Lady Spencer-Churchill, the Honourable Lady Sarah Audley and the Honourable Lady Soames. Reprinted by permission of Cassell Ltd., Houghton Mifflin Company, and the Canadian publisher, McClelland and Stewart Ltd., Toronto.

Colette: from *The Evening Star*, trans. David le Vay. Copyright © 1974 by Colette. Reprinted by permission of Peter Owen Ltd., Publishers, and The Bobbs-Merrill Company, Inc.

Cyril Connolly: from *Enemies of Promise*. Copyright 1983 by Deirdre Levi. Reprinted by permission of Persea Books, New York.

Noel Coward: 'Middle Age' from *The Girl Who Came to Supper*. Reprinted by permission of Dr Jan Van Loewen Ltd; from *Remembered Laughter: The Life of Noel Coward* by Cole Lesley. Copyright © 1976 by Cole Lesley. Reprinted by permission of the Cole Lesley Estate, Jonathan Cape Ltd., and Alfred A. Knopf, Inc.

Gordon Craig: from *Gordon Craig: The Story of His Life* by Edward Craig (Gollancz, 1968). Reprinted by permission of London Management.

Dostoevsky: from *The Brothers Karamazov*, trans. David Magarshack. Copyright © 1982 David Magarshack; from *The Idiot*, trans. David Magarshack. Copyright © 1970 David Magarshack; and from *Notes from the Underground*, trans. J. Coulson. Copyright © 1972 J. Coulson. All reprinted by permission of Penguin Books Ltd.

Isadora Duncan: from *My Life*. Copyright 1927 by Boni & Liveright. Copyright renewed 1955 by Liveright Publishing Corporation. Reprinted by permission.

Albert Einstein: from *Einstein: The Life and Times* by Ronald Clark (originally published by Macmillan, Inc; new edition, Avon and Harry N. Abrams Inc.). © 1969, 1984 by Ronald Clark. Reprinted by permission of A. D. Peters & Co., Ltd., and Literistic Ltd.

T. S. Eliot: extract from 'Sweeney Agonistes' from *Collected Poems 1909–1962*. Copyright 1936 by Harcourt Brace Jovanovich, © 1963, 1964 by T. S. Eliot. Reprinted by permission of Faber & Faber Ltd., and Harcourt Brace Jovanovich, Inc.

William Faulkner: from *Selected Letters of William Faulkner*, ed. Joseph Blotner. Copyright © 1977 by Jill Faulkner Summers. Reprinted by permission of Random House Inc.

F. Scott Fitzgerald: from *The Crack-Up*. Copyright 1945 by New Directions Publishing Corporation. Reprinted by permission of The Bodley Head and New Directions Publishing Corp.; from *The Letters of F. Scott Fitzgerald*, ed. Andrew Turnbull. Copyright © 1963 Frances Scott Fitzgerald Lanahan. Reprinted by permission of The Bodley Head and Charles Scribner's Sons; from *Early Success*. Reprinted by permission of The Bodley Head.

Gustave Flaubert: from *The Letters of Gustave Flaubert 1830–1857* (1980), ed. and trans. Francis Steegmuller. Reprinted by permission of Harvard University Press and the editor.

E. M. Forster: extracts from *E. M. Forster: A Life* by P. N. Furbank. Copyright © 1977, 1978 by P. N. Furbank. Reprinted by permission of Secker & Warburg Ltd., and Harcourt Brace Jovanovich, Inc.

Sigmund Freud: from *The Life and Work of Sigmund Freud* by Ernest Jones. Reprinted by permission of The Hogarth Press and Basic Books, Inc.; from 'On the sexual theories of children' (published in the United States in *The Collected Papers of Sigmund Freud*, Vol. II, by Basic Books, Inc.) and extract from *An Outline of Psycho-Analysis* (published in the United States by W. W. Norton & Co., Inc.), both collected in *The Standard Edition of the Complete Psychological Works of Sigmund Freud*, trans. and ed. James Strachey. Reprinted by permission of the Sigmund Freud Copyrights Ltd., The Institute of Psycho-Analysis, The Hogarth Press, Basic Books Inc., and W. W. Norton & Co., Inc.

Robert Frost: 'What Fifty Said' from *The Poetry of Robert Frost*, ed. Edward Connery Lathem. Copyright 1928, © 1969 by Holt, Rinehart & Winston. Copyright © 1956 by Robert Frost. Reprinted by permission of the Estate of Robert Frost, Jonathan Cape Ltd., and of Holt, Rinehart & Winston, Publishers.

Hans Gal: Gustav Mahler to Bruno Walter, trans. Hans Gal; and Haydn to his publishers, trans. Daphne Woodward, both quoted in *The Musician's World: Letters of the Great Composers*, ed. Hans Gal (1965). Reprinted by permission of Thames & Hudson Ltd.

André Gide: from *The Journals of André Gide*, trans. Justin O'Brien. Copyright 1947 by Alfred A. Knopf, Inc. Reprinted by permission of Alfred A. Knopf, Inc., and Secker & Warburg Ltd.

Maurice Goudeket: from *The Delights of Growing Old*, trans. Patrick O'Brien. Copyright © 1966 by Farrar, Straus & Giroux, Inc. Reprinted by permission of Farrar, Straus & Giroux, Inc., and Laurence Pollinger Ltd.

Robert Graves: from 'Fools'; from 'The Imminent Seventies' and extract from the Introduction to *The Collected Poems of Robert Graves* (Cassell, 1975). Reprinted by permission of Robert Graves and A. P. Watt Ltd.

Graham Greene: from *A Sort of Life*. Copyright © 1971 by Graham Greene. Reprinted by permission of Laurence Pollinger Ltd., and Simon & Schuster, Inc.

Edward Hanslick: from *Vienna's Golden Years of Music 1850–1900*, trans. and ed. Henry Pleasants III. Reprinted by permission of John Farquharson Ltd.

Edward Heath: from *Sailing*. Reprinted by permission of Sidgwick & Jackson.

Ernest Hemingway: from *Selected Letters 1917–1961*, ed. Carlos Baker. Copyright © 1981 The Ernest Hemingway Foundation, Inc.; copyright © Carlos Baker.

Reprinted with the permission of Granada Publishing Ltd., and Charles Scribner's Sons.

A. P. Herbert: 'There was an old man who said "Damn . . ."'. Reprinted by permission of Lady Gwendolyn Herbert and A. P. Watt Ltd.

Ronald Hingley: from *Joseph Stalin: Man and Legend*. Reprinted by permission of A. D. Peters & Co., Ltd.

Julian Huxley: from *Memories 1970–1973*. Reprinted by permission of George Allen & Unwin (Publishers) Ltd., and Harper & Row Publishers, Inc.

Pope John XXIII: from *Journal of a Soul*, trans. Dorothy White. Copyright © 1965, 1980 by Geoffrey Chapman, a division of Cassell Ltd. Reprinted by permission of Doubleday & Company Inc., and Geoffrey Chapman.

Ernest Jones: from *Free Associations: Memories of a Psycho-analyst*. Reprinted by permission of the author's Literary Estate, The Hogarth Press and Basic Books, Inc.

James Joyce: from *A Portrait of the Artist as a Young Man*. Copyright 1916 by B. W. Huebsch, copyright 1944 by Nora Joyce. Copyright © 1964 by the Estate of James Joyce. Reprinted by permission of The Society of Authors as the literary representative of the Estate of James Joyce, Jonathan Cape Ltd., and Viking Penguin, Inc; from *Ulysses*. Copyright 1914, 1918 by Margaret Caroline Anderson and renewed 1942, 1946 by Nora Joseph Joyce. Reprinted by permission of The Bodley Head and Random House, Inc.

Carl Jung: from *The Collected Works of C. G. Jung*, trans. R. F. C. Hull, Bollingen Series XX, Vol. 17: *The Development of Personality*. Reprinted by permission of Routledge & Kegan Paul Ltd., and Princeton University Press.

Rose Fitzgerald Kennedy: from *Times to Remember*. Copyright © 1974 by The Joseph P. Kennedy Jr. Foundation. Reprinted by permission of Doubleday & Company, Inc.

Hugh Kingsmill: from *The Dawn's Delay* (Elkin Matthews, 1924). Reprinted by permission of Curtis Brown Ltd., London.

Rudyard Kipling: 'The Waster'. Copyright Caroline Kipling; extracts from 'Back to the Army Again' and 'My Rival', all from *The Definitive Edition of Rudyard Kipling's Verse*. Reprinted by permission of A. P. Watt Ltd. on behalf of The National Trust, Macmillan, London, Ltd., and Doubleday & Company, Inc. 'Back to the Army Again' first appeared in *Barrack Room Ballads*, published by Methuen & Co.

Oskar Kokoschka: from *My Life*. Copyright © 1974 by Thames & Hudson Ltd. Reprinted by permission of Macmillan Publishing Co., Inc.

Philip Larkin: from *The Listener*, 17 August 1972. By permission of the author.

Rosamond Lehmann: from *Invitation to the Waltz* (n.e. 1980). Reprinted by permission of Collins Publishers.

Sinclair Lewis: from *Babbitt*. Copyright 1922 by Harcourt Brace Jovanovich, Inc. Renewed 1950 by Sinclair Lewis. Reprinted by permission of the Estate of Sinclair Lewis, Jonathan Cape Ltd., and Harcourt Brace Jovanovich, Inc.

Herbert Lottman: from *Albert Camus: A Biography*. Reprinted by permission of George Weidenfeld & Nicolson Ltd., and the author.

Louis XIV: from *Louis XIV* by Joanna Richardson. Reprinted by permission of George Weidenfeld & Nicolson Ltd.

Robert Lowell: 'Sound Mind, Sound Body' from *Notebook*. Copyright © 1967, 1968, 1969, 1970 by Robert Lowell; excerpt from 'Middle Age' from *For the Union Dead*. Copyright © 1962, 1964 by Robert Lowell; excerpts from 'For Sheridan', 'Our Afterlife I (for Peter Taylor)' and 'Art of the Possible', all from *Day By Day*. Copyright © 1975, 1976, 1977 by Robert Lowell. All reprinted by permission of

Farrar, Straus & Giroux, Inc., and Faber & Faber Ltd.
Alison Lurie: from *The War Between the Tates*. Copyright © 1974 by Alison Lurie. Reprinted by permission of William Heinemann Ltd., and Random House, Inc.
Carson McCullers: from *The Member of the Wedding* (Cresset, 1947). Reprinted by permission of Hutchinson Publishing Group.
Phyllis McGinley: 'The Velvet Hand'. Copyright 1951 by Phyllis McGinley, renewed © 1979 by Julie Elizabeth Hayden and Phyllis Hayden Blake; 'A Certain Age'. Copyright 1956 by Phyllis McGinley, renewed © 1984 by Phyllis Hayden Blake. Originally published in the *New Yorker*, both from *Times Three*. Reprinted by permission of Secker & Warburg Ltd., and Viking Penguin, Inc.
Sir Compton Mackenzie: from *My Life and Times: Octave 10, 1953–1963* (Chatto, 1971). Reprinted by permission of The Society of Authors as the literary representative of the Estate of Sir Compton Mackenzie.
Norman Mailer: from *Armies of the Night*. Copyright © 1968 by Norman Mailer. Reprinted by arrangement with New American Library, New York, and Scott Meredith Literary Agency.
Thomas Mann: from *Diaries 1918–1939*, trans. Richard and Clare Winston, ed. Hermann Kesten. © S. Fischer Verlag GmbH, Frankfurt am Main, West Germany 1977, 1978, 1979, 1980. English translation © 1982 Harry N. Abrams Inc. All rights reserved. Reprinted by permission of André Deutsch and Harry N. Abrams, Inc., New York; from *Buddenbrooks*, trans. H. T. Lowe-Porter. Copyright 1924 and renewed 1952 by Alfred A. Knopf, Inc. Reprinted by permission of Secker & Warburg Ltd., and Alfred A. Knopf Inc.
Don Marquis: from *O Rare Don Marquis* by Edward Anthony. Copyright © 1962 by Paul Reynolds. Reprinted by permission of Doubleday & Company, Inc., and Paul Reynolds, Inc.
Guy de Maupassant: from *An Old Man*, trans. Francis Steegmuller and published in *Maupassant: A Lion In the Path*. Copyright 1949 by Francis Steegmuller, copyright © renewed by Francis Steegmuller, 1977. Reprinted by permission of McIntosh & Otis, Inc., and Macmillan, London & Basingstoke.
Vladimir Mayakovsky: from 'The Cloud in Trousers' from *The Bed Bug and Selected Poetry*, trans. Max Hayward and George Reavey (Indiana University Press, 1975). Reprinted by permission of Laurence Pollinger Ltd.
Henry Miller: from *Sextet*. Copyright © 1977 by Henry Miller. Reprinted by permission of Capra Press, Santa Barbara, California.
A. A. Milne: extract from 'Disobedience' in *When We Were Very Young*. Copyright 1924 by E. P. Dutton Co., Inc; renewed 1952 by A. A. Milne. Reprinted by permission of Methuen Children's Books, E. P. Dutton, New York, and McClelland & Stewart Ltd., Toronto. From *Now We Are Six*. Reprinted by permission of Methuen Children's Books.
Dr Maria Montessori: from *The Absorbent Mind* (Kalakshetra Publications, Madras, India, 1959).
Grandma Moses: in *A Tribute to Grandma Moses*. Copyright © 1955, Grandma Moses Properties Co., New York.
Iris Murdoch: from *The Bell*. Copyright © 1958 by Iris Murdoch. Reprinted by permission of Viking Penguin Inc., and Chatto & Windus Ltd.
Dr Margaret Murray: from *My First Hundred Years*. Reprinted by permission of William Kimber & Co., Ltd.
Vladimir Nabokov: from *Lolita* and *Speak, Memory: An Autobiography Revisited*. Reprinted by permission of George Weidenfeld & Nicolson Ltd.
Ogden Nash: 'Lines on Facing Forty', copyright 1942 by Ogden Nash; 'Calendar Watchers, or What's so Wonderful About Being a Patriarch', copyright 1952 by

Ogden Nash and first appeared in the *New Yorker*. Both from *Verses From 1929 On* and reprinted by permission of Curtis Brown Ltd, London, on behalf of the estate of Ogden Nash, and Little, Brown & Company: 'You Can Be a Republican, I'm a Gerontocrat', copyright 1950 by Ogden Nash, first appeared in the *New Yorker*; 'Let's Not Climb the Washington Monument Tonight', copyright 1947 by the Curtis Publishing Company; 'Not George Washington, Not Abraham Lincoln's, But Mine', copyright 1940 by The Curtis Publishing Company; 'Eheu! Fugaces, or What a Difference a Lot of Days Make', copyright 1949 by Ogden Nash; all three first appeared in the *Saturday Evening Post*. These poems are published in the UK in *I Wouldn't Have Missed It* and in the USA in *Verses From 1929 On*, and are reprinted by permission of André Deutsch and Little, Brown & Company.

Harold Nicolson: from *Diaries and Letters 1930–1964*, ed. S. Olsen. Reprinted by permission of Collins Publishers.

Anaïs Nin: from *Linotte: The Early Diary of Anaïs Nin 1914–1920*. Copyright © 1978 by Rupert Pole as Trustee under the Last Will and Testament of Anaïs Nin; from *The Diary of Anaïs Nin 1931–1934*. Copyright © 1966 by Anaïs Nin. Reprinted by permission of Peter Owen Ltd., Publishers, and Harcourt Brace Jovanovich, Inc.

George Orwell: from 'Extracts from a Manuscript Notebook' in *The Collected Essays, Journalism and Letters of George Orwell*, Vol. 4. Copyright © 1968 by Sonia Brownell Orwell; from 'Why I Write' in *Such, Such Were the Joys*. Copyright 1953 by Sonia Brownell Orwell, renewed 1981 by Mrs George K. Perutz, Mrs Miriam Gross & Dr Michael Dickson, Executors of the Estate of Sonia Brownell Orwell; from 'Such, Such Were the Joys' in *Such, Such Were the Joys*. Copyright 1952, 1980 by Sonia Brownell Orwell. Reprinted by permission of A. M. Heath & Co., Ltd., on behalf of the Estate of the late Sonia Brownell Orwell and Martin Secker & Warburg Ltd., and of Harcourt Brace Jovanovich, Inc.

Dorothy Parker: 'Ballade at Thirty-five' from *The Portable Dorothy Parker*. Copyright 1926, copyright renewed 1954 by Dorothy Parker. Reprinted by permission of Viking Penguin, Inc.: published in the UK in *The Collected Dorothy Parker* and reproduced by permission of Gerald Duckworth & Co., Ltd.

Helene Parmelin: from *Picasso Says*, trans. Christine Trollope. Reprinted by permission of George Allen & Unwin (Publishers) Ltd.

Henri Perruchot: from *Gaugin*, trans. Humphrey Hare. Reprinted by permission of Piccadilly Rare Books and Perpetua.

Jean Piaget: from *Judgement and Reasoning in the Child* and from *The Child's Conception of Physical Causality*. Reprinted by permission of Routledge & Kegan Paul Ltd., and the US publisher, Humanities Press Inc.; from *Six Psychological Studies*, trans. Anita Tenzer. Reprinted by permission of Hodder & Stoughton Ltd. (formerly University of London Press).

Sylvia Plath: excerpt from 'Lady Lazarus' from *The Collected Poems of Sylvia Plath*, ed. Ted Hughes. Copyright © 1963 by Ted Hughes. Also in *Ariel*. Copyright © 1965 Ted Hughes. Reprinted by permission of Olwyn Hughes and Harper & Row, Publishers Inc.

Plato: from *The Republic*, trans. Desmond Lee (Penguin Classics, 2nd edn. revised 1974). Copyright © H. D. P. Lee, 1955, 1974. Reprinted by permission of Penguin Books Ltd.

Katherine Anne Porter: from *Katherine Anne Porter: A Life* by Joan Givner. Copyright © 1982 by Joan Givner. Reprinted by permission of Jonathan Cape Ltd., for the author, and Simon & Schuster, Inc.

John Cowper Powys: from *Letters to Nicholas Ross* and from *The Art of Growing Old*. Reprinted by permission of Laurence Pollinger Ltd., and the Estate of John

Cowper Powys.

John Prickett: from *Initiation Rites*. Reprinted by permission of Lutterworth Press.

V. S. Pritchett: from *Midnight Oil*. Reprinted by permission of Chatto & Windus Ltd., for the author, and A. D. Peters & Co., Ltd.

Marcel Proust: extracts from *Remembrance of Things Past*, Vol. III: *The Captive, The Fugitive*, and *Time Regained*, trans. Terence Kilmartin and Andreas Mayor. Translation © 1981 by Random House, Inc., and Chatto & Windus. Reprinted by permission of the publishers; from *Marcel Proust: A Biography*, Vol. II by George D. Painter. Copyright © 1965 by George D. Painter. Reprinted by permission of Chatto & Windus Ltd., and Random House, Inc.

Jean Renoir: from *Renoir, My Father*, trans. Randolph and Dorothy Weaver. Reprinted by permission of the William Morris Agency, New York, and A. D. Peters & Co., Ltd.

Arthur Rimbaud: 'Ma Bohème' and 'The Seven-Year-Old Poet', both from *The Drunken Boat*, trans. Brian Hill.

John D. Rockefeller: from *John D. Rockefeller* by Allan Nevins. Copyright 1940 Charles Scribner's Sons, copyright renewed 1968 Allan Nevins. Reprinted by permission of Charles Scribner's Sons.

Irving Rosenwater: from *Sir Donald Bradman* (B. T. Batsford, 1978). Reprinted by permission of the publisher.

Philip Roth: from *Portnoy's Complaint*. Copyright © 1967, 1968, 1969 by Philip Roth. Reprinted by permission of Jonathan Cape Ltd., for the author, and Random House, Inc.

Artur Rubinstein: from *My Many Years* and *My Young Years*. Reprinted by permission of Irving Paul Lazar and Jonathan Cape Ltd.

Bertrand Russell: from *Dear Bertrand Russell*, ed. Feinberg and Kasrils. Copyright © 1969 by George Allen & Unwin Ltd. Reprinted by permission of George Allen & Unwin (Publishers) Ltd., and Houghton Mifflin Company; from *The Autobiography of Bertrand Russell*, Vols. 1, 2 and 3. Reprinted by permission of George Allen & Unwin (Publishers) Ltd.,; from *On Education*. Reprinted by permission of George Allen & Unwin Ltd., and W. W. Norton & Co., Inc., as US publisher.

Saint-Saëns: from *Saint-Saëns and His Critics* by James Harding (Chapman & Hall, 1963). Reprinted by permission of Associated Book Publishers Ltd.

George Sand: from *My Life*, trans. Dan Hofstadter. Reprinted by permission of The Balkin Agency, Dan Hofstadter's agent.

Carl Sandburg: from *Abraham Lincoln: The Prairie Years and The War Years*. Copyright 1926 by Harcourt Brace Jovanovich, Inc. Renewed 1954 by Carl Sandburg. Reprinted by permission of the publisher.

Jean-Paul Sartre: from *The Words*, trans. Bernard Frechtman. Reprinted by permission of Hamish Hamilton Ltd., and Georges Braziller, Inc.

Sir George Schuster: from *Private Work and Public Causes: A Personal Record 1881–1978*. Reprinted by permission of D. Brown & Sons.

Anne Sexton: from *Writers at Work*, The Paris Review Interviews, 4th series, ed. George Plimpton. Copyright © 1974, 1975 by the Paris Review Inc. Reprinted by permission of Secker & Warburg Ltd., and Viking Penguin, Inc; 'Hurry up please, it's time' from *The Death Notebooks*. Copyright © 1974 by Anne Sexton; 'Old' from 'In a dream you are never eighty' from *Selected Poems*. Copyright 1962 by Anne Sexton. Reprinted by permission of The Sterling Lord Agency.

Bernard Shaw: from *The Revolutionist's Handbook*, *Farfetched Fables*, and Bernard Shaw to Hesketh Pearson, June 1950. Reprinted by permission of The Society of Authors on behalf of the Bernard Shaw Estate.

R. J. Shepherd: from *The Fit Athlete*. Copyright © OUP 1978. Reprinted by permission of the publisher.

Lord Shinwell: from *Lead With the Left: My First Ninety-Six Years*. Reprinted by permission of Cassell Ltd.

Sibelius: from *Sibelius. A Personal Portrait* by Santeri Levas (1972). Reprinted by permission of J. M. Dent & Sons Ltd., and Curtis Brown Ltd.

Sir Osbert Sitwell: from *Tales My Father Taught Me* (Hutchinson, 1979) and *Laughter in the Next Room*: Autobiography Vol. 4, (Macmillan, 1949). Reprinted by permission of David Higham Associates Ltd.

Stevie Smith: 'The Conventionalist' and 'Papa Love Baby' from *The Collected Poems of Stevie Smith*. Copyright 1972 by Stevie Smith. Reprinted by permission of James MacGibbon and New Directions Publishing Corporation; extract from 'Childhood and Interruption' from *Me Again: Uncollected Writings of Stevie Smith*. Copyright © James MacGibbon 1937, 1938, 1942, 1950, 1957, 1962, 1966, 1971, 1972, 1981. Reprinted by permission of Virago Ltd., and Farrar, Straus & Giroux, Inc.

C. P. Snow: from *Variety of Men*. Reprinted by permission of Curtis Brown Ltd., London on behalf of the Estate of C. P. Snow, and Macmillan, London & Basingstoke.

Theodore C. Sorensen: from *Kennedy*. Copyright © 1965 by Theodore C. Sorensen. Reprinted by permission of Harper & Row, Publishers, Inc., and Hodder & Stoughton, Ltd.

Benjamin Spock: from *Baby and Child Care*. Copyright © 1945, 1946, 1957, 1968, 1976 by Benjamin Spock, MD. Reprinted by permission of Pocket Books, a division of Simon & Schuster, Inc., and The Bodley Head.

Enid Starkie: from *Arthur Rimbaud*. Copyright © 1961 by Enid Starkie. Reprinted by permission of New Directions Publishing Corporation, and Faber & Faber Ltd.; from *Baudelaire*. Reprinted by permission of Faber & Faber Ltd.

Richard Strauss: from *The Correspondence between Richard Strauss and Hugo Von Hofmannsthal*, trans. Hans Hammelmann and Ewald Osers. Reprinted by permission of Collins Publishers.

Tchaikovsky: from *Tchaikovsky: A Self-Portrait* by Vladimir Volkoff. Reprinted by permission of Robert Hale Ltd.

Dylan Thomas: 'Twenty-four years remind the tears of my eyes . . .' from *Collected Poems* (Dent, 1952). Reprinted by permission of David Higham Associates Ltd.

James Thurber: from *Selected Letters of James Thurber*, ed. Helen Thurber and Edward Week; from 'Preface to a Life'. Copyright 1933, © 1961 James Thurber, in *My Life and Hard Times*, published by Harper & Row. Reprinted by permission of Hamish Hamilton Ltd., and Mrs James Thurber.

Leo Tolstoy: from *Childhood, Boyhood, Youth*, trans. Rosemary Edmonds (Penguin Classics, 1964). Copyright © Rosemary Edmonds, 1964. Reprinted by permission of Penguin Books Ltd.; from *Tolstoy's Letters*, ed. and trans. R. F. Christian. © R. F. Christian 1978. Reprinted by permission of The Athlone Press; from *Tolstoy* by Henri Troyat, trans. Nancy Amphoux. Reprinted by permission of W. H. Allen & Co., Ltd.

Toscanini: from *Toscanini* by Harvey Sachs. Reprinted by permission of George Weidenfeld & Nicolson Ltd.

Robert C. Twombly: from *Frank Lloyd Wright*. Reprinted by permission of John Wiley & Sons Ltd.

Vincent Van Gogh: from *Dear Theo: The Autobiographies of Vincent Van Gogh*, ed. and trans. Irving Stone, Copyright 1937 by Irving Stone. Reprinted by permission of Doubleday & Company, Inc.

Gore Vidal: from *Rocking the Boat*. Copyright © Gore Vidal 1963. Reprinted by permission of Curtis Brown Ltd., London and William Heinemann Ltd.

Arthur Waley: 'On Being Sixty' by Po Chü-I from *The Life and Times of Po Chü-I*. Reprinted by permission of George Allen & Unwin (Publishers) Ltd.

Evelyn Waugh: from *The Letters of Evelyn Waugh*, ed. Mark Amory. Copyright © the Estate of Laura Waugh 1980. Copyright © in the Introduction and compilation Mark Amory 1980; from *The Diaries of Evelyn Waugh*, ed. Michael Davie. Reprinted by permission of George Weidenfeld & Nicolson Ltd., and Ticknor & Fields, a Houghton Mifflin Company.

Tennessee Williams: 'Kitchen Door Blues' from *In the Winter Cities*. Copyright © 1946 by G. Schirmer. Copyright © 1956 by Tennessee Williams. Reprinted by permission of Elaine Greene Ltd., Literary Agency, and of New Directions Publishing Corporation.

Edmund Wilson: excerpts from *Upstate*. Copyright © 1971 by Edmund Wilson. Reprinted by permission of Farrar, Straus & Giroux, Inc.

P. G. Wodehouse: from *My Man Jeeves*; from Preface to *The Golf Omnibus* and from *P. G. Wodehouse: Portrait of a Master* by David A. Jasen. Reprinted by permission of Scott Meredith Literary Agency and A. P. Watt Ltd.

Virginia Woolf: from *A Writer's Diary*. Copyright 1953, 1954 by Leonard Woolf. Renewed 1981, 1982 by Quentin Bell and Angelica Garnett; from *The Diary of Virginia Woolf*, Vol. III 1925–1930. Copyright © 1980 by Quentin Bell and Angelica Garnett; from *The Letters of Virginia Woolf*, Vol. VI, 1936–1941. Copyright © 1980 by Quentin Bell and Angelica Garnett; and from *Mrs Dalloway*. Copyright 1925 by Harcourt Brace Jovanovich, Inc. Renewed 1953 by Leonard Woolf. All reprinted by permission of The Hogarth Press, the author's Literary Estate and Harcourt Brace Jovanovich, Inc.

W. B. Yeats: 'You think it horrible that Lust and Rage' from *Letters on Poetry from W. B. Yeats to Dorothy Wellesley* (OUP, 1940). Reprinted by permission of A. P. Watt Ltd., on behalf of Michael B. Yeats and Anne Yeats; from *The Letters*, ed. Allan Wade. Copyright 1953, 1954, and renewed 1982 by Anne Butler Yeats. Reprinted by permission of A. P. Watt Ltd., on behalf of Michael B. Yeats, and Macmillan Publishing Co; 'Why Should Not Old Men Be Mad?' (from 'On the Boiler'). Copyright 1940 by Georgie Yeats; renewed 1968 by Bertha Georgie Yeats, Michael Butler Yeats and Anne Yeats; from 'The Tower'. Copyright 1928 by Macmillan Publishing Co., Inc., reenewed 1956 by Georgie Yeats; 'Lines Written in Dejection' (from *The Wild Swans at Coole*). Copyright 1919 by Macmillan Publishing Co., Inc., renewed 1947 by Bertha Georgie Yeats; 'Pardon, Old Fathers' (from *Responsibilities*). Copyright 1916 by Macmillan Publishing Co., Inc., renewed 1944 by Bertha Georgie Yeats; all from *The Poems*, ed. Richard J. Finneran (published in the UK as *The Collected Poems*). Reprinted by permission of A. P. Watt Ltd., and Macmillan Publishing Co.

While every effort has been made to secure permission, we may have failed in a few cases to trace the copyright holder. We apologize for any apparent negligence.

INDEX OF NAMES

Adams, Henry, 109
Adams, John, 78, 113, 161, 168
Adams, John Quincy, 113, 145, 168
Adenauer, Konrad, 172
Alcott, Louisa M., 40
Alice, Princess, Countess of Athlone, 180
Allen, Woody, 43
Amis, Kingsley, 15, 102
Andersen, Hans, 132
Angerville, Mouffle D', 35
Anon., 18, 19, 23, 28, 36–7, 58, 89, 111, 122, 135, 160, 176, 182
Apple, R. W., 180
Aristotle, 2, 13, 19, 43, 99
Arnold, Matthew, 84, 90
Arnold, Thomas, 102
Ascham, Roger, 43
Ashford, Daisy, 88
Asquith, Herbert H., 122
Astor, Lady, 106
Aubrey, John, 29
Auden, W. H., 40, 125, 127
Augustine, Saint, 2, 43, 156
Austen, Jane, 39, 45, 49, 55, 60, 65, 80, 86
Avery, Peter, 135

Bacall, Lauren, 60
Bailey, Paul, 81
Baily, Leslie, 114
Balzac, Honoré de, 67, 76, 98
Barnard, Dr, 97
Barrie, J. M., 65, 106
Baruch, Bernard M., 162, 168
Bashkirtseff, Marie, 46
Baudelaire, Charles, 51
Bax, Clifford, 84
Beaton, Cecil, 133, 144, 160
Beauvoir, Simone de, 40, 107, 123, 127
Beckett, Samuel, 42
Beerbohm, Max, 23, 55, 159, 175
Beethoven, Ludwig van, 12, 26
Belloc, Hilaire, 18, 30, 33, 60
Bellow, Saul, 116

Bennett, Arnold, 67
Berenson, Bernard, 155, 161, 167, 173, 175
Berger, John, 128
Berkeley, Bishop George, 129
Betjeman, Sir John, 57
Bettelheim, Bruno, 47
Birren, James E., 93
Blackstone, Sir William, 21
Blake, Eubie, 183
Blake, William, 3
Blythe, Ronald, 120, 174
Bonnard, Pierre, 127
Borrow, George, 58
Boswell, James, 3, 77, 124
Bowen, E. E., 114
Boyle, Andrew, 144
Bradman, Sir Donald, 82
Bramwell-Booth, Catherine, 179
Brantôme, Pierre de, 37
Briand, Aristide, 83
Brittain, Vera, 46
Brockway, Lord (Fenner), 173
Brontë, Charlotte, 39, 69
Brooke, Rupert, 50, 60
Browne, Sir Thomas, 64, 71
Browning, Elizabeth Barrett, 5, 26
Browning, Robert, 21
Burgon, J. W., 174
Burns, Robert, 51
Burton, Robert, 47, 135
Bury, J. B., 20
Butler, Samuel, 75
Buxton, Frank, 179
Byron, Lord (George Gordon), 67, 71, 73, 75, 76, 78, 117

Camus, Albert, 66, 85
Carlyle, Thomas, 8, 56, 63, 71, 79
Carroll, Lewis (C. L. Dodgson), 20
Casals, Pablo, 153, 176
Casanova de Seingalt, G. G., 95
Castiglione, Count B., 2
Cato, 153, 158
Cellini, Benvenuto, 47, 81, 113

Cervantes, Miguel de, 99
Cézanne, Paul, 132
Chamfort, Nicholas, 178
Channon, Chips, 86
Chaucer, Geoffrey, 19, 31, 85, 117, 153
Chekhov, Anton, 53, 61, 96
Chesterfield, Lord (P. D. Stanhope), 21, 41
Chesterton, G. K., 10
Chichester, Sir Francis, 121, 138
Christie, Agatha, 144
Churchill, Lady Randolph (Jennie), 54, 124
Churchill, Sir Winston, 32, 52, 54, 125, 127, 143, 145, 148, 160
Cicero, 94, 120, 158, 182
Clark, Ronald W., 180
Claudel, Paul, 151
Clement, 156
Coleridge, Samuel Taylor, 17, 22, 58, 75
Colette, 123
Columbus, Christopher, 89
Connolly, Cyril, 75
Conrad, Joseph, 25, 92
Constable, John, 111
Cooper, Lady Diana, 164, 168
Cornaro, Luigi, 158, 177, 184
Corot, Jean, 137, 146
Coward, Noël, 93, 111, 128
Cowley, Abraham, 29
Cowley, Malcolm, 149
Crabbe, George, 105, 146
Craig, Gordon, 175, 178

Dandolo, Enrico, Doge of Venice, 180
Dante Alighieri, 24, 56, 72, 74, 154
Darwin, Charles, 4, 7
Davis, Colin, 80
Deffand, Madame du, 140
Degas, H. G. E., 58
Delacroix, Eugène, 80, 98, 107
De la Mare, Walter, 175
Denning, Lord, 155
Desmoulins, Camille, 71
Dickens, Charles, 5, 24, 27, 31, 46, 51, 55
Disraeli, Benjamin, 79, 140, 149, 152
Dodgson, C. L., see Carroll, Lewis
Donne, John, 24
Dostoyevsky, F. M., 22, 85, 107, 111
Drennan, Robert E., 136
Dryden, John, 38, 131

Dufferin, Lady (Helen Selina), 148
Duncan, Isadora, 3
Dunne, F. P., 64
Durkheim, Émile, 43

Edgeworth, Maria, and R. L., 11
Einstein, Albert, 33, 56
Eliot, George, 24, 61, 178
Eliot, T. S., 1, 102, 104
Emerson, Ralph Waldo, 62, 68, 77, 87, 112, 113, 118, 122, 135, 168, 180
Erlanger, Philippe, 13
Evelyn, John, and Dick, 7

Faulkner, William, 121
Fielding, Henry, 69
Fielding, Sir John, 32
Fitzgerald, Scott, 41, 46, 52, 61, 65, 81, 102, 127
Flaubert, Gustave, 44, 70
Fontenelle, Bernard de, 178, 182
Ford, F. Madox, 15
Forster, E. M., 107, 129, 157
Frank, Anne, 37
Franklin, Benjamin, 26, 49, 69, 83, 88, 126, 147
Frederick the Great, 63
Freland, Mrs, 182
Freud, Sigmund, 11, 26, 109, 127, 143, 153
Frost, Robert, 106

Gandhi, Mahatma, 35
Gaskell, Elizabeth, 39
Gauguin, E. H. P., 74
Gay, John, 126
Gesell, Arnold, 14, 26
Gibbon, Edward, 22, 103, 108
Gide, André, 77, 88, 105, 110, 124, 129, 142, 146, 150
Gilbert, W. S., 89, 96, 114
Gladstone, William Ewart, 17, 115, 157, 159, 163, 164
Goethe, J. W. von, 12, 75, 142, 143, 153, 154, 156
Gogol, Nikolai, 62
Golding, William, 33
Goldsmith, Oliver, 102, 139
Goudeket, Maurice, 125, 144
Goya, F. J. de, 151-2
Graves, Robert, 92, 134, 150
Green, Julien, 18
Greene, Graham, 35, 50, 120, 141

Greville, Charles, 128
Grey, Edward (of Fallodon), 143
Guermantes, Duc de, 159

Haggard, D. John, 163
Hanslick, Eduard, 121
Hardy, Thomas, 84, 102, 112, 136, 141, 156, 163, 165
Haydn, Joseph, 16, 129, 140
Hazlitt, William, 91
Heath, Edward, 100
Heath-Stubbs, J., 135
Heine, Heinrich, 143
Hemingway, Ernest, 91, 115
Heraclitus, 181
Herbert, A. P., 155
Herbert, George, 49
Herodotus, 156
Herzen, Alexander, 31, 42
Hitler, Adolf, 30
Hobbes, Thomas, 170, 172
Hodges, C. Walter, 178
Hoffmannsthal, Hugo von, 108
Hokusai, 143
Holland, Lady, 21
Holmes, Oliver Wendell, 142, 172
Holmes, Oliver Wendell, jun., 171, 174, 183
Homer, 156
Hood, Thomas, 9, 95
Horowitz, Vladimir, 147
Housman, A. E., 53
Howells, William, 131
Hubbard, Kin, 115
Hudson, W. H., 23
Hugo, Jules, and Leopold, 108
Hugo, Victor, 83, 108
Hume, David, 126
Hummel, J. N., 12
Huxley, Julian, 13

Ibsen, Henrik, 93
Inge, W. R., 166
Isherwood, Christopher, 48, 137
Isocrates, 181

Jacob, Max, 58
James, Henry, 25, 59
Jefferson, Thomas, 89, 112, 113, 139, 145
Jesus Christ, 71, 89, 154
'JM', 105
Joan of Arc, 34

John XXIII, Pope, 150
Johnson, Samuel, 3, 13, 45, 51, 68, 74, 94, 96, 97, 99, 106, 124, 130, 137, 140, 143
Jones, Ernest, 18
Jonson, Ben, 19
Joyce, James, 42, 70
Jung, C. G., 6, 145, 162

Kallir, Otto, 184
Karloff, Boris, 25
Keats, John, 54, 75
Kennedy, J. F., 89
Kennedy, Rose F., 157
Keynes, Sir Geoffrey, 177
Kingsley, Charles, 37
Kingsmill, Hugh, 53
Kipling, Rudyard, 54, 57, 99, 136
Klein, Melanie, 6
Knight, Joseph, 171
Knox, Ronald, 13
Kokoschka, Oskar, 172
Kreisler, Fritz, 10

La Bruyère, Jean de, 1, 52
Lamb, Charles, 10, 27, 66
Lampedusa, Giuseppe de, 141
Landor, Walter Savage, 75, 152, 168
Lang, Andrew, 86
Larkin, Philip, 103
Laski, Harold J., 183
Lawrence, D. H., 36, 42, 61, 66, 84, 88, 146
Leaf, Andrew, 181
Léautaud, Paul, 77
Lehmann, R. C., 87, 110
Lehmann, Rosamond, 44
Lennon, John, 124
Leslie, Anita, 121
Lewis, C. S., 116
Lewis, Sinclair, 95
Lewis, Wyndham, 79
Lincoln, Abraham, 105, 111
Lindsay, John V., 120
Liszt, Franz, 121
Lloyd George, David, 8
Locke, John, 2, 41
Lockhart, J. G., 34
Longfellow, H. W., 142, 153
Lottman, Herbert, 85
Louis XIV, 13, 130, 144
Louis XV, 35
Lowell, J. R., 84, 130

Lowell, Robert, 29, 67, 92, 101, 103
Lurie, Alison, 38

MacArthur, Douglas, 138
Macaulay, Thomas Babington, 12
McCartney, Paul, 124
McCullers, Carson, 33
McGinley, Phyllis, 17, 36, 137
Mackenzie, Compton, 44, 165
Macklin, Charles, 171
Maclean, Hector, 137
Macmillan, Harold, 157, 164
Madison, James, 89, 113
Mahler, Gustav, 98
Mailer, Norman, 91
Mann, Thomas, 90, 114, 123
Mansfield, Katherine, 68, 73
Marie Antoinette, 73
Marquis, Don, 161
Martin, George, 149
Martin, R. B., 104
Martin, W. K., 167
Marvell, Andrew, 39
Marx, Karl, 48
Mary, Queen of Scots, 37
Maugham, Somerset, 174
Maupassant, Guy de, 168
Mayakovsky, Vladimir, 53
Mayer, Sir Robert, 164, 184
Melbourne, Lord, 119
Meynell, Alice, 32
Michelangelo Buonarroti, 37
Mill, John Stuart, 22, 111
Miller, Henry, 72, 87, 151, 184
Milne, A. A., 9, 17
Milton, John, 29, 56, 75
Miró, Joan, 162
Monnet, Jean, 164
Monroe, James, 113
Montagu, Lady Mary Wortley, 86, 131
Montessori, Maria, 6, 16
Moore, George, 64, 124
More, Hannah, 12
Mortimer, Raymond, 144
Moses, Grandma, 179, 184
Mosley, Sir Oswald, 54
Mozart, Wolfgang Amadeus, 12, 39,
  63, 69, 110
Murdoch, Iris, 55
Murray, James, 6
Murray, K. M. Elisabeth, 6
Murray, Dr Margaret, 185
Musset, Alfred de, 46

Mussolini, Benito, 93

Nabokov, Vladimir, 1, 25
Napoleon Bonaparte, 30, 64
Nash, Ogden, 45, 79, 85, 92, 99, 119
Nevins, Allan, 180
Newman, Arnold, 120
Newman, J. H., 40
Newton, George, 176
Newton, Sir Isaac, 53
Nicolson, Harold, 123
Nightingale, Florence, 57
Nin, Anaïs, 28, 63

O'Casey, Sean, 154
Omar Khayyám, 135
Origen, 156
Orwell, George, 16, 21, 101
Osler, Sir William, 59, 118

Palgrave, F. T., 104
Palmerston, Lord, 149, 152
Parker, Dorothy, 75, 136
Pavese, Cesare, 68, 87
Peacham, Henry, 2
Pearson, Hesketh, 167
Péguy, Charles, 31, 83
Penn, William, 29
Pepys, Samuel, 71, 76
Perruchot, Henri, 74
Philip II, of Spain, 59
Phillpotts, Eden, 175
Piaget, Jean, 7, 16, 19, 41
Picasso, Pablo, 18, 118, 120, 157, 160,
  165-6
Pinero, Sir Arthur Wing, 94
Piozzi, Mrs, 154
Pitt, William, Lord Chatham, 70
Pitt, William, younger, 20-1
Pius IX, Pope, 108
Plath, Sylvia, 68
Plato, 109, 156
Po Chü-i, 117
Pope, Alexander, 20, 154
Porson, Richard, 102
Porter, Katherine Anne, 75
Porter, Peter, 84
Pougy, Liane de, 57
Pound, Ezra, 40
Powys, J. Cowper, 119, 129, 139, 161
Praed, W. M., 62
Prickett, John, 36
Priestley, J. B., 143

Prior, Matthew, 15
Pritchett, V. S., 52, 136
Proust, Marcel, 78, 143, 159
Puccini, Giacomo, 39
Pythagoras, 156, 181

Radiguet, Raymond, 58
Raphael (Raffaello Sanzio), 75, 78, 79
Reith, Lord, 144, 148
Renard, Jules, 83
Renard, Simon, 60
Renoir, Auguste, 137
Renoir, Jean, 138
Reynolds, Frances, 94, 97
Rhodes, Cecil, 98
Rhys, Jean, 50
Rimbaud, Arthur, 20, 42, 58
Ripley, Samuel, 76
Robb, Barbara, 145
Rochester, Earl of, 72
Rockefeller, J. D., 162, 180
Rogers, Samuel, 148
Rooney, Mickey, 38
Rosenwater, Irving, 82
Rossetti, D. G., 24
Rossetti, Olive, 15
Roth, Philip, 72
Rothenstein, William, 152
Rothschild, Lord, 59
Rousseau, Jean Jacques, 17, 31, 41, 45, 103
Routh, Dr Martin, 174, 182
Rubinstein, Artur, 8, 100, 155, 174
Ruskin, John, 15, 89
Russell, Bertrand, 3, 16, 30, 62, 115, 160, 165, 171, 180, 181

Sagan, Françoise, 45, 89
Saint-Évremond, Charles de, 109
Saint-Exupéry, Antoine, 78
St John, Henry (Lord Bolingbroke), 70
Saint-Saëns, C. C., 161, 164
Salinger, J. D., 44
Sand, George, 5, 44
Sandberg, Carl, 105
Sartre, Jean-Pierre, 13, 63
Sassoon, Siegfried, 54
Schiller, J. C. F. von, 49
Schopenhauer, Arthur, 26, 81, 119
Schubert, Franz, 61
Schumann, R. A., 49
Schuster, Sir George, 181
Schweitzer, Albert, 80

Scott, Sir Walter, 34, 118
Sévigné, Madame de, 123
Sexton, Anne, 63, 91, 151
Shackleton, Sir Ernest, 97
Shakespeare, William, 34, 47, 49, 55, 80, 105, 117, 151
Shaw, George Bernard, 64, 85, 167, 173, 178
Shelley, Mary, 65
Shelley, Percy Bysshe, 65, 73, 75, 92
Shepherd, R. J., 86
Sheridan, Richard Brinsley, 43, 76, 106
Shinwell, Lord (Manny), 179
Sibelius, Jean, 159, 166, 169, 172
Sillitoe, Alan, 48
Simonides, 153
Sitwell, Sir George, 147
Sitwell, Sir Oswald, 79, 147
Skinner, John, 163
Smith, Stevie, 4, 10, 38
Smith, Sydney, 139
Snow, C. P., 56
Socrates, 156
Solon, 87
Sophocles, 153, 156, 170
Sorensen, Theodore, 89
Southey, Robert, 8, 50, 162
Spender, Stephen, 45
Spock, Benjamin, 6, 11
Stalin, Joseph, 103
Starkie, Enid, 51, 59
Steele, Richard, 12
Stein, Gertrude, 60
Stendhal (Marie Beyle), 49
Stephen, Thoby, 52
Stern, Isaac, 30
Stevenson, Robert Louis, 69, 83, 124
Storr, Anthony, 3
Strachey, Lytton, 52
Strauss, Richard, 108
Stravinsky, Igor, 120, 166
Strindberg, August, 98
Suetonius, 71
Swift, Jonathan, 70, 73, 87, 96, 111, 112, 120, 170
Swinburne, Algernon Charles, 64
Sydney-Turner, Saxon, 52

Taine, Hippolyte, 4, 7
Talleyrand-Périgord, C. M. de, 159
Tarkington, Booth, 28
Tasso, Torquato, 47
Taylor, A. J. P., 104, 145

Taylor, Jeremy, 37
Tchaikovsky, Peter Ilich, 91, 96, 103
Temple, Shirley, 18
Temple, Sir William, 70
Tennyson, Lord (Alfred), 43, 58, 104, 157
Tennyson, Sir Charles, 167
Tennyson, Hallam, 167
Terence, 41
Thackeray, William Makepeace, 51, 96, 102, 131
Thatcher, Margaret, 75
Theophrastus, 153
Thom, Douglas A., 10
Thomas, Denis, 18, 166
Thomas, Dylan, 57
Thorndike, Sybil, 151
Thucydides, 156
Thurber, James, 57, 81, 117, 123, 175
Tolstoy, Leo, 5, 17, 32, 69, 115, 132, 148, 152
Tolstoy, Sofia, 128
Tomlinson, H. M., 175
Toscanini, Arturo, 165
Traherne, Thomas, 1
Travers, Ben, 169, 177
Trelawny, E. J., 73
Trevelyan, G. O., 12
Trollope, Anthony, 57, 60, 89, 102, 109
Turgenev, Ivan S., 67, 74
Twain, Mark, 47, 84, 88, 121, 138, 140
Twombly, Robert C., 173

Ustinov, Peter, 8

Valéry, Paul, 14
Van Gogh, Vincent, 48, 67
Vasari, Giorgio, 37, 64, 78
Verdi, Giuseppe, 84, 140, 148, 149, 160
Victoria, Queen, 3, 45, 50, 119, 154
Vidal, Gore, 36
Villon, François, 64

Voltaire, F. M. Arouet de, 115, 148, 156

Wagner, Richard, 109
Waley, Arthur, 117
Walpole, Horace, 56, 123, 128, 140
Walton, Izaak, 24
Washington, George, 89, 113
Watson, James D., 59
Waugh, Evelyn, 39, 81, 104, 118
Wayne, John, 107
Weideger, Paula, 31
Weil, Simone, 38
Wellington, Duke of, 156
Wesley, Charles, 125
Wesley, John, 142, 146, 158, 162
West, Rebecca, 158
Wharton, Edith, 94
Whitehead, A. N., 35
Whitman, Walt, 132
Wilde, Oscar, 65, 77, 95
Williams, Tennessee, 172
Wilson, Edmund, 127, 131, 138, 141, 144
Wilson, Harriette, 39
Winnicott, D. W., 10
Wodehouse, P. G., 112, 133, 175, 176
Wolfe, Thomas, 29
Wollstonecraft, Mary, 66
Wood, Anthony à, 172
Woolf, Virginia, 52, 80, 94, 96, 104, 110
Wordsworth, William, 9, 15, 27, 34, 58, 73, 75
Wright, Frank Lloyd, 173

Yeats, William Butler, 98, 101, 118, 119, 123, 137, 139
Yonge, Charlotte M., 9
Young, Edward, 1, 66, 83

Zola, Émile, 121
Zukov, Adolph, 181
Zweig, Stefan, 73